MOLECULAR PHYSICS

MAXWELL PRESS

MOLECULAR PHYSICS

James Arnold Crowther, M.A.

Fellow of St.John's College, Cambridge, and Demonstrator in Physics at the Cavendish Laboratory

MAXWELL PRESS

Chennai Trichy NewDelhi

MAXWELL PRESS

This edition has been published in india by arrangement with Carson Books, UK

ISBN 978-93-90063-58-1

Maxwell Press
No. 44, Nallathambi Street,
Triplicane, Chennai 600 005

MJP 904

Publisher's Note

The legacy of a country is in its varied cultural heritage, historical literature, developments in the field of economy and science. The top nations in the world are competing in the field of science, economy and literature. This vast legacy has to be conserved and documented so that it can be bestowed to the future generation. The knowledge of this legacy is slowly getting perished in the present generation due to lack of documentation.

Keeping this in mind, the concern with retrospective acquiring of rare books has been accented recently by the burgeoning reprint industry. Maxwell Press is gratified to retrieve the rare collections with a view to bring back those books that were landmarks in their time.

In this effort, a series of rare books would be republished under the banner, "Maxwell Press". The books in the reprint series have been carefully selected for their contemporary usefulness as well as their historical importance within the intellectual. We reconstruct the book with slight enhancements made for better presentation, without affecting the contents of the original edition.

Most of the works selected for republishing covers a huge range of subjects, from history to anthropology. We believe this reprint edition will be a service to the numerous researchers and practitioners active in this fascinating field. We allow readers to experience the wonder of peering into a scholarly work of the highest order and seminal significance.

MAXWELL PRESS

PREFACE

THE present volume is an attempt to give a connected account of the constitution and properties of the atom and molecule in the light which has been thrown upon these particles by recent physical research. From the great mass of experimental observations which constitute what has been called somewhat vaguely the "New Physics" I have endeavoured to select those portions which have a direct bearing on this important and fascinating subject, and I have drawn freely on all available sources of information, some of which are indicated in the Bibliography at the end of the book.

To Professor Sir J. J. Thomson I owe a debt of gratitude for his kindness in allowing me to select from his experimental negatives, and to reproduce, the Positive Ray photographs which illustrate Chapter IV. I owe him still more for the privilege which I have enjoyed during the past eight years in the Cavendish Laboratory, of moving in daily contact with much of the work, and of drinking in, as it were at the fountain head, much of the theory narrated in the present volume. To this must be ascribed any merits which the following pages may be found to possess; for their errors and deficiencies I, alone, am responsible.

J. A. C.

CONTENTS

MOLECULAR PHYSICS

CHAPTER I.

INTRODUCTION.

THE history of the rise of molecular theory has often been written. From Dalton, who first gave to the conceptions of the early atomists a definite meaning and a real experimental basis, through Avogadro and his fertile hypothesis of the molecule, and so on to the kinetic theory of Clausius and Maxwell, the story is told in almost every text-book of Chemistry and Physics, and is familiar as household words to all their readers. Here for a time science seemed to pause in her progress. It seemed as if the memoirs of these brilliant mathematicians were to contain the last word in molecular theory.

If this had been so, the following pages would have been superfluous. In the last few years, however, new fields have been opened to the student of molecular phenomena; new methods of approach have been evolved so potent in character that subjects about which we had until then barely sufficient grounds for speculation are now laid open to direct experimental attack. The molecule has been raised from a conception only realisable experimentally in millions to the rank of a definite particle whose entry into our apparatus produces a definite and measurable effect. At the same time the accuracy, and what is still more important, the certainty, of our measurements of molecular magnitudes have been enormously increased. It seemed, therefore, neither undesirable nor devoid of interest to attempt to re-write the subject of molecular physics, not from the standpoint of the kinetic theory, by which it was approached historically, and which constituted for many years the sole line of attack, but rather in the light thrown upon these problems by the recent developments in what has come to be known as the "New Physics."

In order to fix our attention upon the problems to be dealt with, and to recall to our minds some clear ideas of the magnitudes with which we shall have to deal, it may be well to consider briefly the position to which we had been brought by the exponents of the kinetic theory.

The atom itself was a purely chemical conception. It took its rise in an attempt to explain the laws of chemical combination, and represents the smallest mass of an element which can take part in a chemical reaction. By its first exponents it was regarded as homogeneous and indivisible. This view is no longer held. We shall see later that we must conceive of the atom as something approaching a planetary system on an infinitesimal scale. A study of the phenomena of radioactivity has driven us to believe that in them we are dealing with a true decomposition of a true chemical atom; that the element radium is in fact a decomposition product of the atom of uranium, and is itself in the course of transition into an atom of lead, with the emission during the process of several atoms of helium. This break up of an atom, however, is a process which we are permitted to watch, but not to control. So far we have not succeeded in either accelerating or retarding its progress in any way whatever.

In compounds the atoms of the different elements

were supposed to be grouped together into similar particles or molecules, the ratio of the masses of the components being the same in each individual molecule as in the compound as a whole. It was soon found necessary to assume that even in the elements the atoms were usually connected into groups of two or more to form a larger particle, which, unless taking part in a chemical reaction, always moved and acted as a single system. This extension of the theory was due to Avogadro, and his additional hypothesis that the number of these molecules in a given volume of gas under similar conditions is independent of the nature of the gas was the first great generalisation in the physics of the molecule.

It was soon realised that these molecules were particles exceedingly small in size. Various attempts to determine the approximate diameter of the molecules all led to the result that it was not very much greater than 2×10^{-8} cms. Various illustrations have been employed to assist us to grasp the extreme minuteness of this quantity. If, for example, a drop of water were to be magnified to the size of the earth the molecules in it would be of the size of footballs. Or again, as a recent writer has suggested, we may say that the carbon atom in the printed page subtends at the reader's eye the same angle as would be subtended there by a man on the moon.

The number of molecules in any visible portion of matter is, of course, correspondingly large. According to the latest determinations the number of molecules in 1 cub. cm. of air under normal conditions (0° C. and 760 mm. pressure) is $2{\cdot}7 \times 10^{19}$. It is almost hopeless to attempt to conceive this inconceivable swarm of particles. It may, perhaps, be of some assistance to point out that in the highest attainable vacua, when the pressure of the gas is not more than $\frac{1}{10000}$ mm. of mercury, the number of molecules present in every cubic millimetre of the space still exceeds 2,000,000,000.

In a solid body we must regard the molecules as relatively fixed. They are indeed in rapid vibration about their mean position, such vibration constituting the phenomenon which we know as heat. The fact that ancient gems and coins have come down to us through many ages still preserving their sharp outlines and their beautiful and clear-cut engraving shows that, in the absence of external forces, the defection of a molecule in a solid from its original station must be of exceedingly rare occurrence.

In liquids, however, we must regard the molecules as free to move. A liquid has no definite shape of its own, and two liquids if left quite undisturbed will gradually diffuse through each other with a velocity which is in general small but quite measurable. The different molecules possess sufficient energy to escape from the attraction of neighbouring systems, but move only a very short distance through the thickly-crowded space before becoming entangled in other clusters.

Gases differ from liquids in the fact that the space occupied by a given number of molecules is very much greater; and a molecule therefore travels much further before coming into collision with other molecules. Thus a gramme of water substance when transformed into steam occupies 1,600 times the volume which it had in the liquid state. The molecules are therefore $\sqrt[3]{1{,}600}$ or, say, 12 times as far apart in the gas as in the liquid. The actual volume of the molecules themselves is thus only a very small fraction of the volume filled by the gas, the molecules occupying the space between their boundaries not by filling it with their bulk, but by moving about in it in all directions with considerable velocities. The tendency of a gas to spread through the whole of the space accessible to it thus represents the tendency of each molecule to proceed on its path in a straight line when not opposed by any obstacle, while the pressure on the boundaries of the gaseous space is due to the exceedingly rapid series of impacts made on them by the innumerable swarm of molecules in motion.

The velocity with which the molecules in a gas are moving can easily be calculated when the pressure and density are known. The mean velocity of a molecule of oxygen at the freezing point of water is about 425 metres per second, or about 1,000 miles per hour. It is, therefore, somewhat greater than the velocity of sound through air under the same conditions (331 metres per second). In addition to collisions upon the boundaries each molecule collides from time to time with other molecules of the gas, the number of collisions made by a single molecule under ordinary conditions of temperature and pressure being about 6,000,000,000 every second.

The distance travelled by the particle between two collisions is known as its free path. Its actual value between any two collisions depends, of course, upon chance. Its mean value can, however, be calculated from the viscosity of the gas. For air under ordinary conditions it is about $7 \cdot 6 \times 10^{-6}$ cms., or roughly $\frac{3}{1000000}$ of an inch.

This mean free path, although a giant among molecular magnitudes, is still rather smaller than the smallest distance we may ever hope to perceive. Abbe has shown that two objects would have to be separated by rather more than twice this distance before they could be resolved by the best possible microscope working under the best conditions.

The mass of each molecule is, of course, exceedingly minute; according to the recent determinations of Rutherford, which we shall describe later, the mass of a hydrogen atom is $1 \cdot 61 \times 10^{-24}$ gms., a number so minute that it can hardly be said to convey to the mind any impression at all. This number, it will be seen, represents the value in grammes of the chemical unit of atomic weight; the mass in grammes of any atom or molecule can thus be calculated from a table of atomic weights by merely multiplying the chemical atomic or molecular weight by this factor. Small as is this quantity we shall see that its value is known with certainty to within a very few per cent.

Of the quantities with which we have been dealing, the kinetic theory gave accurately the velocity of the molecule and the mean free path. It threw only a dim and somewhat uncertain light upon the number and mass of the individual molecules. As to the nature of the molecule it can hardly be said to have given any information at all. It is at once the strength and the weakness of the statistical methods employed in the kinetic theory, that given a sufficiently large number of molecules, their exact number, size, mass and properties become for many purposes matters of indifference. Thus many of the properties of gases were successfully deduced, in the first place, on the assumption that the molecule was a hard sphere, an assumption which in the case, say, of a molecule like that of oxygen built up of two atoms is almost certainly incorrect. It will be seen then how little light can be thrown upon the individual molecule by investigations of this nature.

As the molecule itself lay far beyond the limits of vision, and as the kinetic theory could afford so little certain information, it might have seemed that we must rest content with the knowledge already attained. Speculation as to the nature of the atom was not wanting, ranging from the simple hard atom of the original theory to the centre of force of Boscovitch, and the vortex atom of Professor

Thomson, the early speculation of one who has done, perhaps, more than any other to render possible a true theory of the atom. These theories, ingenious as they were, withered for lack of experimental support, and the subject became gradually to be regarded as being in the realm of the unknowable. Meanwhile the great success of the statistical methods of the kinetic theory, the extension of thermodynamics, and the study of phenomena from a consideration of their equilibrium conditions which followed on the work of Willard Gibbs, were leading further and further away from atomic and molecular hypotheses.

Fortunately one loophole remained into this region of the infinitely small. The kinetic energy of a particle is equal to the product of one-half its mass into the square of its velocity. However small the mass of the particle, therefore, we can theoretically by sufficiently increasing its velocity endow it with an appreciable amount of energy. In the case of a molecule the velocity required will, of course, be very large. The energy of a molecule in the air under normal conditions of temperature and pressure is about 5×10^{-14} ergs, assuming the mass and velocity already given. This is far too small to be detected. Let us suppose, however, that we can endow our molecule with a velocity much greater than this, approaching 2×10^9 cms. per second. The energy of an oxygen molecule moving with such a speed would be $\frac{1}{2}(32 \times 1{\cdot}61 \times 10^{-24}) \, . \, (2 \times 10^9)^2$ ergs, or about 10^{-4} ergs. This, though small, is not less than can be appreciated in other forms of energy. Thus Lord Rayleigh has estimated that the faintest sound which can be heard transmits every second across a square centimetre of area at right angles to its path about $2{\cdot}3 \times 10^{-5}$ ergs. The energy in the faintest perceptible beam of light is not far different from this. Thus the energy of impact of a single molecule travelling with the velocity we have suggested would be comparable with the energy in a faint beam of light or a just audible sound.

The velocity ascribed to the molecule in this calculation might seem impossibly large. It is in fact about $\frac{1}{15}$ of the velocity of light, and a particle travelling with this speed would, therefore, cover the distance from the earth to the sun in 2 hours. Such particles, however, actually exist, and it is the discovery of these particles and the measurements made upon them that have led to the great advances in molecular physics which we are about to describe. Particles having this velocity are shot out in large numbers from radioactive bodies. To anticipate a little we may say that the α-particles from radium consist of atoms of helium shot out with a speed of this order of magnitude, and bearing a positive charge. Thus it is that a single α-particle is able to cause a flash of light when it strikes upon a screen covered with a suitable material.

The α-particles consist of helium atoms only. Velocities approaching that of the α-particles can be given to atoms and molecules of other substances by passing an electric discharge through them in the gaseous state at very low pressures. The phenomena of the discharge tube have indeed afforded the best means of investigating the properties of moving electrified particles, and we shall proceed to their consideration immediately.

In order to complete our survey we will anticipate so far as to say that the first experimental investigation of these phenomena led to results of a startling nature. Professor Sir J. J. Thomson, as the result of a brilliant series of researches and deductions, which may truly be said to mark a fresh era in the progress of science, showed that the most conspicuous set of particles in the discharge tube, namely, those constituting the cathode rays, had a mass not more than $\frac{1}{1800}$ part of that of a hydrogen atom.

It was found, moreover, that whatever the nature of the gas through which the discharge was passing, whatever the nature of the electrodes used, the mass of the cathode particle produced and the magnitude of the charge carried by it was always the same. These cathode particles, corpuscles or electrons, as they came to be called, must thus be regarded as forming an integral part of the atom of every kind of substance. They are therefore even more fundamental than the atom itself, since they form part of the material from which the atom is built up. The methods used in their investigation are similar in principle to those which have recently been applied to atoms and molecules, but the special properties of the electrons rendered their experimental investigation less difficult than that of the atom or molecule, so that for a time it could truly be said that far more was known of the electron than of the atom itself. For these reasons we will commence our study of molecular physics with the consideration of the electron.

CHAPTER II.

THE PHYSICS OF THE ELECTRON.

THE general appearance of an ordinary discharge tube is perhaps familiar to most of my readers. It is represented somewhat diagrammatically in Fig. 1. Covering the cathode C is a velvety light known as the cathode glow. Beyond this is a comparatively dark space, named after its discoverer Sir William Crookes. If the pressure in the tube is fairly high this Crookes dark space will be so close to the cathode and so small as to be hardly visible. Beyond

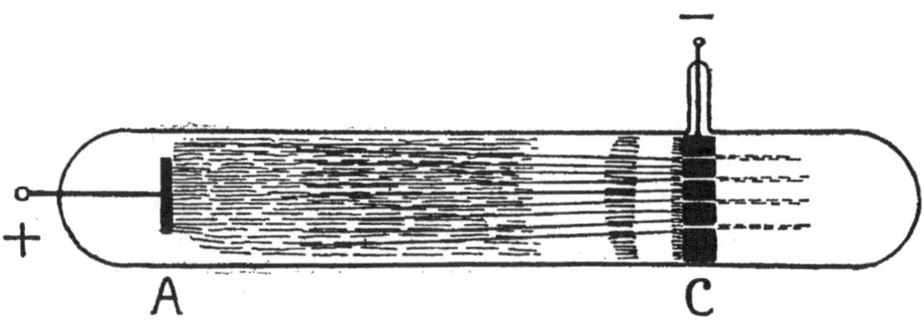

FIG. 1.—THE DISCHARGE TUBE.

this dark space is another luminous area, the negative glow, bounded on the side remote from the cathode by still another obscure area, the Faraday dark space. The length of the tube occupied by these phenomena depends mainly on the pressure of the gas, and is practically independent of the size of the tube, the remainder of which, however long, is filled with a bright light known as the positive column, which is frequently, though not necessarily, divided into numerous striæ. It is the light from this positive column that is generally employed in the spectroscopic examination of gases in the discharge tube.

If the pressure is fairly low another phenomenon may be distinguished. Proceeding normally from the cathode and penetrating in straight lines across one or more of the dark spaces may be seen faint bluish streams of light, known as the cathode rays. As the pressure is still further reduced these rays become relatively more and more important, and penetrate further and further, until finally they fall upon the glass walls of the discharge tube, exciting there a greenish phosphorescence. Meanwhile the Crookes dark space has grown until it occupies practically the whole of the tube, the remaining parts of the ordinary discharge being represented by the glow on the cathode and a single patch of light near the anode itself.

The nature of the cathode rays remained for some time in doubt. However, in 1895 Perrin showed that they transported a negative charge, and in the following year Professor Sir J. J. Thomson proved conclusively that they consisted of streams of negatively charged particles, moving with extremely high velocities and having a mass in all probability much less than that of a hydrogen atom.

Let us continue for a moment the study of the discharge tube. It had long been known that if a solid obstacle were placed in the path of the discharge a shadow of the object was cast by the cathode rays on the opposite wall of the tube. It was noticed that if the obstacle was placed within the Crookes dark space a shadow was also cast on the cathode. That is to say, there must also be rays in the tube proceeding in the opposite direction to the cathode rays and falling upon the cathode itself. By using a perforated cathode, as shown in the diagram, these rays become visible as faintly luminous streaks of light, coming from each of the perforations, and exciting a faint mauve phosphorescence where they strike the walls of the tube. These rays which carry a positive charge and have a mass comparable with that of an atom, form the positive particles to which we have already alluded.

The high velocities of these particles give them, as we have seen, sufficient energy to produce luminous effects when they fall upon suitable screens. The additional fact that they carry an electric charge enables us to determine not only the velocity, but also the ratio of the mass of the particles to the charge which they carry.

The general problem of the motion of an electrified particle when moving under the action of electric and magnetic forces is one of some little complexity, though complete solutions have been given by various mathematicians. Fortunately the special cases which are required for our present purpose lend themselves readily to the simplest treatment.

Suppose that a particle of mass m and charge e is moving with a velocity v at right angles to a magnetic field of strength H. In one second the charge e will have moved through a distance v cms. We may therefore regard the path of the particle as a sort of conductor v cms. long, carrying an electric current whose strength is e. The mechanical force on such a conductor when placed in a magnetic field is known from the laws of electro-magnetic induction to be equal to the strength of the field multiplied by the current and the length of the conductor, that is to the product $H.e.v$. Further, this force is at right angles to the direction of the current and to the lines of force in the magnetic field. A little consideration will show that under these circumstances the path of the particle is changed from a straight line to a circle described in a plane at right angles to the direction of the lines of force in the magnetic field.

Let r be the radius of this circle. A particle of mass m moving round a circle with a velocity v exerts a centrifugal force tending to urge it away from the centre of the circle. The value of this is known to be

$$m \cdot v^2/r.$$

Now for equilibrium the force tending to urge the particle away from the centre must equal the force due to the magnetic field tending to draw it in. Thus we have

$$H \cdot e \cdot v = m \cdot v^2/r$$

$$\frac{m}{e} \cdot v = H \cdot r \quad \text{. . . . (a)}$$

Again the particle carries an electric charge. It can therefore be deflected by an electric field. Let X be the strength of the electric field. Then the force experienced by the particle is equal to $X \cdot e$ and acts upon the particle in the direction of the lines of electric force. If now we arrange our apparatus so that the electric field and the magnetic field are at right angles to each other, the forces upon the particle due to the two fields will act along the same straight line, and by properly adjusting the sign of the electric field we can make the two forces oppose each other. It will then be possible to find a value for the electric field such that the two forces just neutralise each other and the particle goes on its path without deflection. When this is so we know that the mechanical forces on the particle due to electric and magnetic fields just balance each other. That is

$$H \cdot e \cdot v = X \cdot e$$

$$v = X/H \quad \text{. . . . (b)}$$

Thus if we apply to the moving electrified particle

simultaneously an electric and magnetic field of such magnitudes that the particle pursues its path without any deviation, the ratio of the strengths of the two fields will give us directly the velocity of the particle. If, further, we can measure the radius of the circle into which the path of our particle is bent by the action of the magnetic field alone, we can, by equation (a), determine the value of the ratio m/e, or the mass to the charge for the particle we are investigating, since the value of v is now known from the previous experiment.

This ratio of the mass to the charge is one of the utmost importance in molecular physics. It may be noted that for ions in solutions it is the quantity of matter associated with the transference of unit quantity of electricity, that is to say, the electro-chemical equivalent. According to the latest values, the electro-chemical equivalent of hydrogen ions in solution is ·0000109 gm. per coulomb. The values of the ratio for other ions can be deduced from this and a table of combining weights in the usual way, since every ion in solution carries by Faraday's laws either the same charge as the hydrogen ion or some simple multiple of it.

Owing to the way in which the volt and the ohm have been defined, the former as 10^8 and the latter as 10^9 absolute electro-magnetic units, the absolute unit of electro-magnetic charge is ten times the coulomb or practical unit of charge. The value of m/e for hydrogen in absolute units of charge is, therefore, ·000109, or just a little greater than 10^{-4} gms. per absolute electro-magnetic unit of charge.[1]

---

[1] The science of electricity is burdened with two distinct systems of units for the measurement of electrical magnitudes, the one founded on the consideration of the force with which two similar charges repel each other, the other on considerations of the induction of currents by magnets. The practical system of units is founded upon the latter, the volt being 10^8 and the ohm 10^9 absolute electro-magnetic units. The ampere

The principles just explained found their first application in the experiments of Professor Sir J. J. Thomson, on the cathode rays, experiments which may well be said to mark a new era in science. The experimental methods used by him have been

FIG. 2.—CATHODE RAY DEFLECTION TUBE.

modified and improved upon in the light of subsequent discoveries, and the accuracy attained in the measurements greatly increased. In principle, how-

---

and coulomb are thus 1/10th of the absolute units of current and charge. To make confusion worse confounded, it has become usual in works and original papers on these subjects to give the value of m/e in electro-magnetic units, while the value of e itself is usually given in electrostatic units. Our statement that the force on the electron due to the magnetic field is equal to $H.e.v$ implies that we are using the absolute electro-magnetic system of units, as this system has been defined in such a way as to make this statement true. As this system is the one more closely related to the units in actual use, we shall employ it throughout, except where the contrary is explicitly stated.

ever, the experiments remain the same, and on account of its historic interest the original apparatus used in these experiments will be described. Fig. 2 is taken (by kind permission of Professor Sir J. J. Thomson) from a photograph of one of the tubes used by him for the experiment. The arrangement of the various parts will be seen more clearly from the diagrammatic representation in Fig. 2A.

The cathode of the discharge tube is at C, and consists of an aluminium disc supported by a rod of the same metal and connected with the outside of the tube by a platinum wire sealed through the glass. The anode R is an aluminium rod placed in

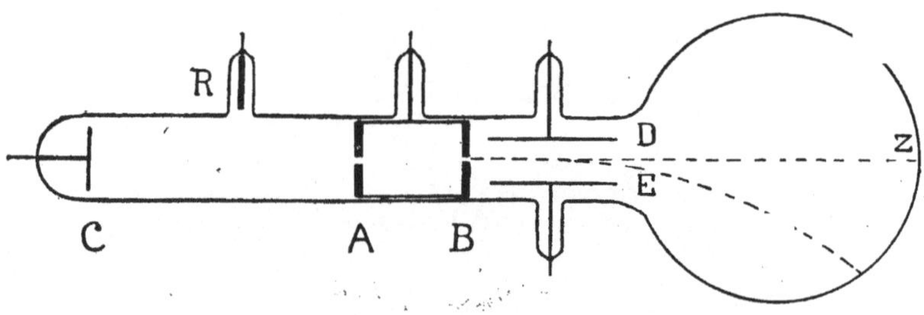

FIG. 2A.—DIAGRAM OF CATHODE RAY TUBE.

a side tube, while facing the cathode is a thick metal disc A, filling the tube but pierced with a small hole through which the cathode rays can pass in a narrow pencil. To define this beam of rays still further, a second metal disc B pierced with a still finer hole is placed some distance behind A. For convenience these two discs are mounted in a metal tube as shown in the figure. This tube and the anode R are connected to earth, while R and C are connected to the terminals of a large induction coil as shown in the photograph.

Thus, when the discharge is passed between C and D, a very fine well-marked pencil of cathode rays emerges from B. It is then made to pass between two parallel plates D and E of aluminium which can be charged to any required difference of potential by connecting them to a large battery of small accumulators.

This part of the tube is also placed between the poles of the little electro-magnet shown in the photograph, the lines of magnetic force being at right angles to the plane of the paper and therefore at right angles to the electric field. Finally, the stream of cathode particles falls upon the walls of the bulb Z which are covered with powdered zinc blende, its presence being marked by a spot of bright green phosphorescent light at the point of impact. The position of the spot in the original experiments was read off on a scale pasted on the outside of the tube.

To perform the experiment the magnetic field is first applied and the deflection of the spot of light measured. From this deflection and the dimensions of the different parts of the tube it is possible to calculate the radius of the circle into which the path of the rays has been bent by the magnetic field. The deflected and undeflected paths of the particles are indicated by the dotted lines in the diagram. The electric field is then applied and the difference of potential between the two plates gradually increased until the spot of light is brought back to its original position. The ratio of the electric to the magnetic field is then equal to the velocity of the cathode particles. The value of the ratio m/e can be calculated from this velocity and the known magnetic deflection as explained above.

The experiments were not without difficulty. It was at first found very difficult to obtain any deflection of the particles by means of an electric field. This was found to be due to the fact that the passage of the cathode rays through the residual gas in the discharge tube rendered it more or less conducting, so that the particles were moving through what was practically a closed conductor and were thus hardly affected by the external electric field. It was necessary in order to eliminate this effect to work with very high vacua.

On making the calculations it was found that the velocity of the particles depended principally upon the potential of the discharge, and to some extent also on the pressure of the residual gas in the tube and the nature of the electrodes and the gas in the tube. Even in the same tube the velocity of all the rays was not the same, so that on applying the magnetic field the single spot of light made by the cathode beam on the zinc blende screen was spread out into a little magnetic spectrum, as we may call it, each point of which corresponded to cathode rays moving with a definite velocity. In tubes at an exhaustion suitable for these experiments the velocities range, as a rule, from 2×10^9 to 3×10^9 cms. per second, or about 1/10th of the velocity of light.

On evaluating the ratio m/e it was not found to suffer similar variations. The original experiments showed and this has since been confirmed by many subsequent experiments, that, whatever the nature of the gas in the discharge tube, whatever the material of the cathode, whatever the working potential of the tube, the value of the ratio of the mass of the cathode particle to its charge is always the same. According to the latest determinations the value of this constant is $5{\cdot}65 \times 10^{-8}$ gms. per unit of charge.

Again, although we have so far considered only the case of the cathode rays, they are not by any means the only source of these corpuscles, or electrons as they are now called. Similar electrons are given off in quantities from a clean metal surface when illuminated by ultra-violet light, or from the surface of the liquid sodium-potassium alloy under the action of the light from an ordinary candle. Rontgen rays produce them whenever they fall upon any material object, and it is interesting to note that the velocity of the electrons thus produced is the

FIG. 3.—MAGNETIC DEFLECTION OF THE CATHODE RAYS.

same as that of the cathode rays in the Crookes tube from which the Rontgen rays proceed. Electrons are also emitted from incandescent metals at a high temperature, and in very large quantities when certain metallic oxides and notably those of barium and calcium are raised to a red heat. These electrons have very little velocity of their own, but may very readily be given one by acting upon them with an electric field. If the heated oxide is raised to a potential P in absolute units, the work done on the electron by the field will be $P \, . \, e$. If v is the velocity acquired by the electron, its kinetic energy is $\frac{1}{2}mv^2$. Hence, by the principle of the conservation of energy $P \, . \, e = \frac{1}{2}mv^2$ or $v^2 = 2P \, . \, e/m$. Thus by suitably altering the potential of the hot oxide we can give to these electrons a suitable and known velocity.

Fig. 3 is a reproduction of an actual photograph of such a discharge. The cathode is a platinum strip carrying a minute speck of barium oxide, raised to a red heat by means of an electric current and charged to a potential of — 240 volts from a cabinet of small accumulators. The electrons given out by the heated BaO are projected by the field in a narrow pencil and are rendered visible by their action on the residual molecules of air in the exhausted tube. In the photograph a magnet has been placed behind the tube so that there is a magnetic field at right angles to the paths of the electrons, which are thus bent into the arc of a circle as shown in the figure. The fact that the circle is not completed is due to the gradual diffusion and absorption of the electrons by collision with the molecules of the gas in the tube.

The value of the ratio m/e for these electrons has been measured by their discoverer Wehnelt and others. It may be noticed that as the potential P to which the cathode is raised is equal to $\frac{1}{2}v^2 \, . \, m/e$, it is only necessary to measure the radius of the circle into which the rays are bent by a magnetic field of known strength H in order to be able to calculate both v and m/e. This fact has been applied by Lenard and others to measure the value of m/e for the electrons given off by metallic surfaces under the action of ultra-violet light.

TABLE I.

Observer.	Source of electrons.	e/m e.m.u. per gm.
Kaufmann . .	β-rays (slow) from Radium	$1{\cdot}77 \times 10^7$
Simon . .	Cathode rays	$1{\cdot}72 \times 10^7$
Ewers . .	β-rays from Polonium .	$1{\cdot}7 \times 10^7$
Classen . .	Cathode rays	$1{\cdot}774 \times 10^7$
,, . .	Hot lime	$1{\cdot}776 \times 10^7$
Woltz . .	β-rays (slow) from Radium	$1{\cdot}767 \times 10^7$
Bestelmeyer .	Hot lime	$1{\cdot}766 \times 10^7$

The above table (Table I.) gives a few of the more recent determinations of the ratio e/m for electrons from different sources. It will be seen that within very narrow limits indeed, especially considering the difficulties of the experiments, the ratio of the charge to the mass is the same for all electrons, whatever their origin and whatever their velocity, and differs very little from $1{\cdot}77 \times 10^7$. The value of m/e or the electro-chemical equivalent of the electron, if we may so call it, is therefore $5{\cdot}65 \times 10^{-8}$. For the hydrogen atom, as we have seen, the value is about 10^{-4}. If we could assume that the charge e is the same for the electron as for the ion in solution we see at once that the mass of the electron must be only about $\frac{1}{1800}$ part of that of the hydrogen atom.

For this assumption, however, very little direct evidence could be adduced at the time it was made. It was one of those strokes of genius which have proved to be correct. To determine absolutely the mass of an electron we must find some method of determining the value of the electronic charge e. Fortunately this also has proved possible, owing to a discovery made in the Cavendish laboratory by Mr. C. T. R. Wilson.

If a space saturated with water vapour is

suddenly chilled the surplus water over and above that which would saturate the space at the lower temperature in general separates out in little particles of water, forming within the space a thick mist or cloud. It was, however, shown many years ago by Aitken that this condensation always commenced round minute particles of dust, particles often far too small to be visible, yet sufficiently material to be caught by a well-packed tube of cotton wool. In the absence of these particles he found that a very considerable degree of supersaturation could be produced without any deposition of water and without any appearance of cloud or mist, and that it was only when the supersaturation became at least eightfold that a mist or cloud was produced. It appeared then that for a lesser degree of supersaturation than this the water vapour must have some nucleus about which to condense before a drop could be formed.

Gases under ordinary conditions are almost perfect insulators. Under the action of certain agents, such, for example, as Rontgen rays or any of the rays from radio-active substances, a gas becomes more or less conducting. It has been shown that this is due to the creation in the gas of systems, some negatively, some positively charged. In all probability these are produced by the ionising agent expelling a negative electron from the atom through which it passes, leaving the remainder of the molecule positively charged. The negative electron thus set free soon attracts to itself one or more atoms just as an uncharged pith ball is attracted by an electrified ebonite rod. The gas thus becomes filled with oppositely electrified systems consisting of clusters of three or four atoms together, which move under the action of an electric field in exactly the same way as the ions of an electrolyte move in the electrolysis of a solution. There is this difference, however, that, whereas the electrolytic ions are stable, the gaseous ions if left to themselves, gradually re-combine to form neutral molecules. Thus unless the supply is kept up by the continuous action of the ionising agent the conductivity will soon die away.

The mechanism of ionisation by the electrons emitted by a metal plate under the action of ultra-violet light is somewhat different. Here the electrons have not sufficient energy to expel an electron from the atom and thus no positive ions are produced in the gas. The negative electrons, however, attract to themselves neutral molecules to form negative ions. It will be evident from the mode of formation that these ions bear the same charge as the electron which gave rise to them, that is to say, their ionic charge is the same as the electronic charge *e*. It may be pointed out in passing that since the ions are all of the same sign and produced at the same place, namely, the surface of the metal plate, a current can only be sent through the gas in one direction. If the metal plate is negative the ions will be repelled and thus carry a negative current from it to the positive electrode. If, however, the plate is positively charged the ions will be attracted back to it by the field and the gas will remain an insulator.

It occurred to C. T. R. Wilson that these ions might possibly serve as nuclei for the production of a fog or cloud in supersaturated air, and this was found on experiment to be the case. He was able to show that drops of water would condense on these negative ions when the supersaturation was only fourfold, while condensation occurred on the positive ions also when the supersaturation was sixfold. It was found that when the supersaturation was increased to the proper extent each negative ion became the centre of a single drop of water, and if the conditions were fairly uniform these drops were all of the same size. It will be remembered that in the case of the ions formed by ultra-violet light each of these drops will carry with it the electronic charge e, the magnitude of which we wish to determine.

Every drop in a cloud is falling through the air under the action of gravity with a definite velocity. Clouds do not float in the air, though, if the drops are sufficiently small they fall very slowly, owing to the great surface they present in comparison with their mass to the viscous drag of the air. Sir G. G. Stokes has shown that if r is the radius of a small drop of water and q the viscosity of the air, the velocity with which the drop will fall relative to the air is $\frac{2gr^2}{9q}$, where g is, as usual, the acceleration due to gravity. The drops formed round the ions are all of the same size and fall at the same rate. We can thus measure the velocity v of their fall by watching the rate at which the surface of the cloud settles. This will enable us to calculate the radius r of the drop, and hence its mass M.

So far the cloud has been falling freely under the action of gravity. Let us apply a field of strength X in such a direction as to oppose the action of gravity, that is to say, with the positive plate above the cloud. The force on each drop due to the electric field is then $X . e$, while that due to gravity is $M . g$. We can adjust the field X until these two forces exactly balance each other, and the drop remains suspended in air like Mahomet's coffin. When this stage is reached we have

$$X . e = M . g$$

or

$$e = M . g/X.$$

But all these quantities are known, and thus the charge on the drop of water, which we have shown to be the same as that on the original electron, can be determined.

It remains to describe the apparatus devised for

the experiment. It is shown in Fig. 4. It is well known that if a mass of gas is expanded in such a way that no heat is allowed to enter or leave it, the temperature will fall. The lowest temperature reached in any given expansion can be calculated

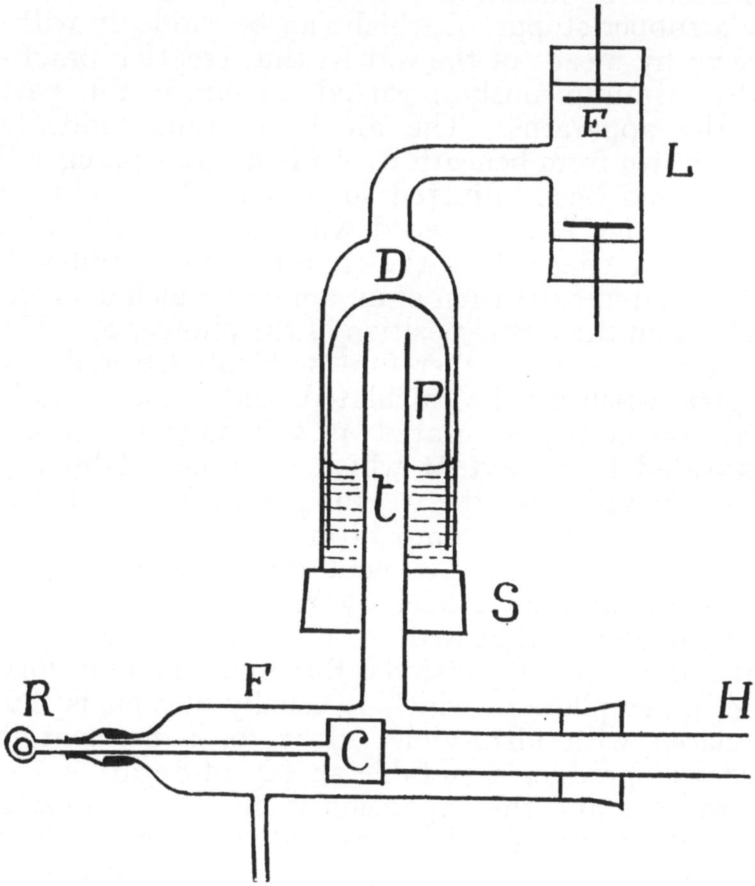

FIG. 4.—C. T. R. WILSON'S EXPANSION APPARATUS.

from the gas laws if the ratio of the specific heats of the gas is known. As gases are bad conductors of heat we can satisfy the conditions of the experiment by making the expansion sufficiently rapid. The mass of the gas is then cooled down to the calculated temperature before any appreciable amount of heat has had time to enter the system.

The method of making the expansion will be clear from the diagram. The tube H connects with a large bulb of some litres capacity kept exhausted by means of a water pump. It is closed by a rubber stopper C which can be suddenly withdrawn by means of the rod R, thus creating practically instantaneously a partial vacuum in this part of the apparatus. The air being thus suddenly withdrawn from beneath it, the light glass plunger P which had been adjusted to a suitable height in the cylinder D is forced down with great rapidity, thus producing within the experimental chamber L a very rapid expansion of an amount which depends only upon the initial position of the plunger P. The temperature in L is therefore suddenly lowered to a degree which can be calculated, and if the air in L was previously saturated it will now be supersaturated to an extent which is measured by P/p where P and p are the vapour pressures of water at the initial and final temperatures.

In making an experiment the air in L is freed from all material nuclei by repeated expansions; the plunger is adjusted so as to make the supersaturation after expansion rather more than fourfold, the upper plate E, which is usually of zinc, is illuminated with ultra-violet light for a moment in order to produce a suitable supply of negative ions in the gas and the expansion is made. A cloud is formed on the negative ions and the rate at which the surface of it settles is taken. After it has fallen some little distance the upper plate E, which had been previously uncharged, is raised to a positive potential, and its potential adjusted until the surface of the cloud is seen to remain stationary. If V is this potential and the lower plate at a distance d from the upper one is earthed, the strength of the electric field is V/d and the electronic charge e is equal to Mgd/V, where M has been calculated by Stokes' law from the rate of fall of the cloud.

The method was first applied in a rather different form by Professor Sir J. J. Thomson. In the form described, which is the one which has been used with the greatest success, it was performed by H. A. Wilson. We have, for the sake of making our argument logically complete, considered the case of the electrons produced by ultra-violet light, for in this case we have evidence that the charge we are measuring is really that carried by an electron. It has been shown, however, experimentally, that the value of the charge e is the same no matter how the ions are produced. Experiments carried out with ions produced by Rontgen rays or the various rays from radio-active materials gave exactly the same values for the charge e. Like the ratio m/e, the charge e is a universal constant. Some recent determinations of it are given in the following table.

TABLE II.

Observer.	Carrier.	e (electrostatic units).
Begeman	Water drops (original method)	$4{\cdot}67 \times 10^{-10}$
Roux	Sulphur drops	$4{\cdot}17 \times 10^{-10}$
Ehrenhaft	Metallic dust	$4{\cdot}65 \times 10^{-10}$
Millikan	Oil or mercury spray	$4{\cdot}89 \times 10^{-10}$
Rutherford	[Counting the α-rays]	$4{\cdot}65 \times 10^{-10}$
Planck	Theory of radiation	$4{\cdot}69 \times 10^{-10}$

It will be seen from Table II. that the values obtained for the electron charge e by different observers working in different ways, though not so consistent as those obtained for the ratio e/m, yet do not differ much from a mean value of about $4 \cdot 7 \times 10^{-10}$ electrostatic units, or $1 \cdot 57 \times 10^{-20}$ in the absolute electromagnetic unit of charge which is equal to 10 coulombs.

The most extensive survey of the subject has been made by Millikan (see *Physical Review*, 1911). Instead of water drops he used very fine globules of oil or sometimes mercury produced by an atomiser or fine spray. These drops could be passed as required into his apparatus, which very much resembled the chamber L in Fig. 4, through a small hole made in the top plate. These globules did not evaporate appreciably, so that an individual drop could be isolated and kept under observation through a microscope for several hours consecutively. The microscope was fitted with an engraved glass scale in the eyepiece, on which the whole of the measurements on the motion of the drop could be made very accurately and conveniently. By properly adjusting the difference of potential between the upper and lower plates the drop could be kept in the field of view during the whole of the experiment. The measurements made and the calculation of the charge are precisely the same for the oil globule as for the water drop.

These drops were not originally charged, but picked up their charges from the surrounding air, which was ionised either by Rontgen rays or by radium. These comparatively large globules could collect and carry more than one electronic charge, seven or eight being a very usual number, though some drops had many more than this. The interchange of charges between the globule and the air could be easily observed with the microscope. The drop which had been poised almost motionless in the air for some time by the balanced action of gravity and the electric field would be seen to dart suddenly upwards, showing that the drop had captured another charge and thus caused the electric force to increase, or perhaps on the other hand the drop would be seen to fall, having parted with one of its charges to some colliding molecule.

The important point brought out by the experiment is that in every case it was found that the magnitude of the charge thus exchanged, which could be calculated by a simple application of the principles already explained, was exactly the electronic charge *e*, and the total charge on the globule at any time was always very exactly some integral multiple of this charge. An examination of the mass of figures collected by Professor Millikan leaves no room for doubting that the charge which we have denoted by *e* is a universal constant, and that every electrical charge is made up of an integral number of electrons, just as every fragment of an element is made up of a whole number of atoms.

From Table I. we have the ratio e/m, equal to $1{\cdot}77 \times 10^7$, for all electrons. Combining this result with the value just obtained for *e* we are driven to the conclusion that whatever the source from which they come all electrons have identically the same mass and charge, the former being $8{\cdot}8 \times 10^{-28}$ gms., the latter $1{\cdot}57 \times 10^{-19}$ coulombs. Such particles must form a constituent part of all matter.

The value of *e* ascribed to Professor Rutherford and also included in Table II. was obtained from experiments of an entirely different nature made on the α-particles. Before passing to these we must refer very briefly to an exceedingly beautiful extension of the method of drops which has recently been brought to perfection by the originator of the experiments.

In the experiments already described, the clouds obtained were fairly uniform, as the ionisation was intense and time was given for the ions formed to diffuse through the body of the gas. Suppose, however, that we allow, say, a single α-particle to cross our expansion chamber and at the same time make an expansion of the air in the chamber by the method

FIG. 5.—THE TRACKS OF THE α-PARTICLES.

already described. The particle passes through the air, ionising it as it passes, and thus leaving behind it a trail of charged ions. Water drops at once form upon these from the now supersaturated air before they have had time either to diffuse into the rest of the gas or to recombine. If the expansion chamber is suitably illuminated these drops appear as brilliant points of light all along the path traversed by the particle. In the case of the *α*-particle the ions are so numerous that the track of the particle becomes visible as a continuous shining streak of light. In the case of an electron where the ionisation is less intense the appearance presented is that of a string of bright beads. We are thus able not merely to detect, but even to follow the path of a single particle of any kind providing it produces ionisation in a gas.

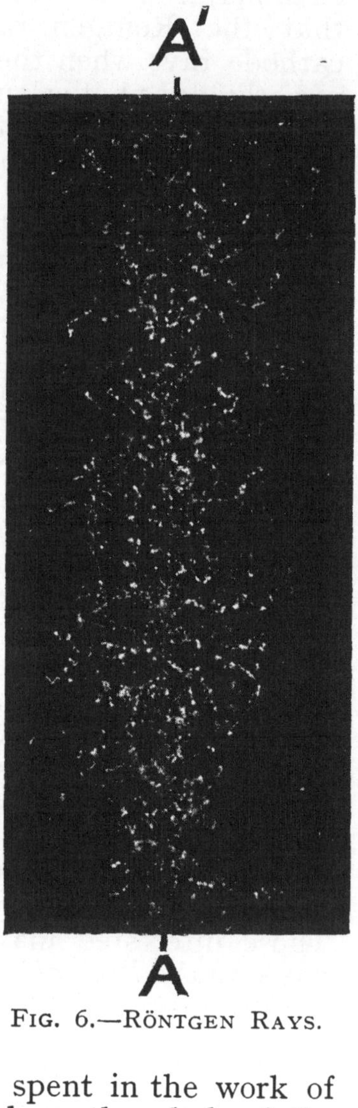

FIG. 6.—RÖNTGEN RAYS.

By the kindness of Mr. C. T. R. Wilson, I am able to reproduce two of the photographs taken by him in this way. In Fig. 5 are shown the paths of a number of *α*-particles. These have their origin in a small sample of radium placed just outside the chamber and shoot across the chamber as shown in the photograph until finally their energy, which is being spent in the work of separating the ionic charges along the whole of the path, finally becomes too small to produce ionisation, and the particles disappear from human ken.

Fig. 6 is a photograph of the passage of a narrow beam of Rontgen rays through the expansion chamber, from A. to A.′ We have mentioned before that the Rontgen rays produce rapidly-moving cathode rays when they fall upon matter; in this remarkable photograph these cathode rays are made visible to us by the ions they produce. Here upon the dark background their tracks may be seen as a tangled skein of bright threads radiating out from the invisible track of the Rontgen rays. Each single thread represents the whole path of a single electron from its expulsion from the parent molecule under the action of the Rontgen rays, to the place where it finally becomes too weak to ionise. In some of the brighter threads the individual drops are clearly to be seen. The twists and turns in each little path are due to collisions between the cathode particle and the molecules of the surrounding air, or, rather, perhaps, with the individual electrons contained in these molecules.

These photographs will perhaps do more than any argument, however cogent, to remove any doubt which may still be felt as to the objective reality of the electrons and other particles with which we have been dealing. Nothing but an actual particle could leave behind it such tracks as those shown in the photographs we have reproduced.

We have not space to deal with the many difficulties of the experiment or with the skill and ingenuity with which they have been overcome. Those interested may be referred to the original paper in the *Proceedings of the Royal Society* for 1912.

The method is still in its infancy, and the photographs so far obtained are hardly more than trial exposures. Already, however, several heated controversies have been settled, and many cherished hypotheses dethroned, as the result of the photographs obtained. To those of us who have been working by indirect methods on some of the more obscure problems still left to us for solution, the method has come as a veritable giving of sight to the blind, and now that Mr. C. T. R. Wilson has perfected his apparatus and can turn his attention to using it as an instrument of research we may hope for many interesting and important results in the near future.

CHAPTER III.

THE POSITIVE PARTICLE.

WE have already seen that positively charged particles are projected with high velocities from radium and other radioactive substances, and that they are also to be found in discharge tubes. Unlike the electron the positive particle has a mass comparable with that of an atom, no positive charge having yet been detected in association with any mass less than that of an hydrogen atom. The charge on a positive particle is, however, either the same as that on an electron or some simple multiple of it. The α-rays, as the particles from radioactive substances are still called (a relic of the time when their nature was unknown) always carry a charge of $2 . e$. In discharge tubes, however, positive particles are found with all integral multiples up to $8 . e$. We shall see that we may regard the positive particles as atoms travelling with a high speed from which one or more electrons have been removed.

The positive particle can be deflected by electric and magnetic fields in the same way as the negative electron, and obeys the same laws. The deflection is in each case in the opposite direction to that of the electron, while, owing to the greater mass, the deflection is less than that for a similarly moving electron. The great ionising power of the positive particles makes the detection of the electrical deflection difficult. For these and other experimental reasons it is only recently that much light has been thrown on their nature and properties. These difficulties have now been overcome, and the positive particle also is beginning to yield up its secrets. As the positive particle is always an atom of some kind, its investigation is, perhaps, from some points of view of even greater interest than that of the electron.

The greater energy of these particles is, however, in some ways an advantage. It enables us to deal with individual particles, whereas in the case of the electron we can only deal with large numbers. Thus our measurements gain, perhaps, in definiteness although they may lose somewhat in numerical precision.

If e be the charge on a single α-particle, the total charge carried by N particles is N . e, and if N is sufficiently large this charge will be large enough to be directly measured by an electroscope or electrometer. If we can further determine the number of particles carrying this charge we shall have made a direct determination of the charge e.

The experiments are best performed using the α-particles from radium, polonium, or some other radioactive substance. They were carried out almost simultaneously by Rutherford, and by Regener. In Rutherford's experiments the α-particle was detected by the ionisation it produced in a special chamber; the entry of each α-particle into the apparatus being notified by a kick of the needle of a sensitive electrometer. In Regener's experiments the number of scintillations produced by the particles in a given time on a fluorescent screen was counted. The former method was the more satisfactory from a theoretical view, as there was at the time no direct evidence that every particle produced a scintillation when it struck a screen, although this has since been shown to be the case by Rutherford. It was also proved by the fact that the two experimenters reached almost exactly the same value for the charge on the α-particle. The experiment of Regener is simpler to carry out than the electrical method of Rutherford, and granting his assumptions, is probably capable of greater accuracy.

Regener's apparatus is shown in Fig. 7. Fig. 7a shows the "counting" apparatus, Fig. 7b the

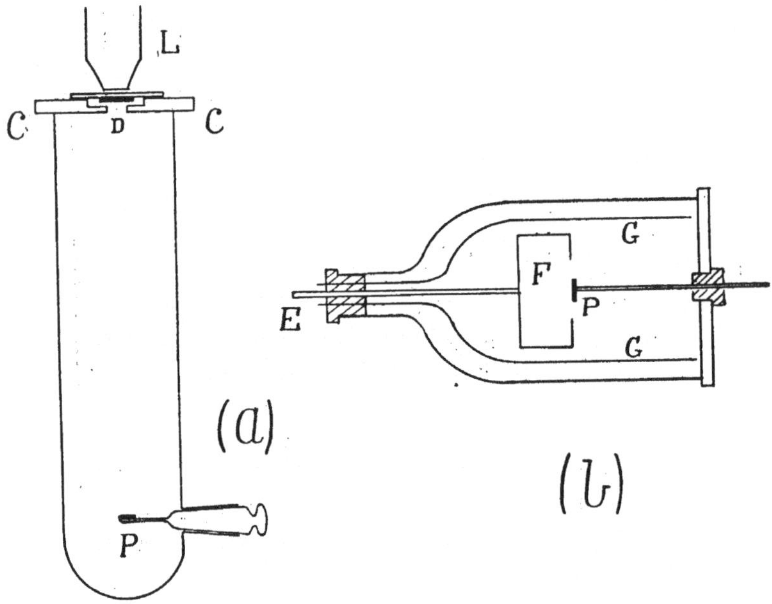

FIG. 7.—REGENER'S APPARATUS.

apparatus for measuring the total charge carried by the rays.

The counting apparatus is very simple. It consists of a long tube closed at one end by a brass plate *c*, *c*, bored with a circular hole some centimetre in diameter. Across this is sealed a thin glass cover-slip bearing on its under surface the fluorescent screen D, in this case a thin plate cut from a crystal of brown diamond, a substance chosen by Regener after many trials as the most suitable for the purpose. Rutherford has since used prepared screens of powdered willemite with equal success.

A high power microscope was focussed on the diamond plate for viewing the scintillations which, numbering about one every two seconds, were easily counted. The source of the α-particles in Regener's experiments was a plate of polonium, placed at P, some 12·7 cms. from the diamond screen. The whole tube was carefully exhausted so that there should be no absorption of the particles before reaching the screen.

The number of scintillations in a given time was counted, and the area A of the aperture in the brass plate and its distance R from the polonium carefully determined. If n is the number of particles hitting the screen in a time t seconds, then the number of particles emitted by the polonium per second across a hemisphere drawn about its upper surface is

$$N = \frac{2\pi R^2}{A} \cdot \frac{n}{t},$$

assuming that the particles are given out equally in all directions. This assumption was verified by Rutherford in a separate experiment. For the disc used by Regener N was found to be $3·935 \times 10^5$.

To measure the total charge carried by the α-particles from the disc a Faraday cylinder was used. This consisted of a partially closed brass cylinder F (Fig. 7b) enclosed in a glass chamber exhausted to a very high vacuum to avoid any loss of charge by ionisation in residual air. The polonium was mounted at P so that the α-particles from its upper surface entered the cylinder. Here they gave up their charges to it, gradually raising it to a positive potential, which could be measured by an electrometer or electroscope connected to the metal electrode E.

This method of measuring small currents by the electrometer is one in every-day use in this class of experimental work and deserves a little further consideration. Let C be the electrical capacity of the Faraday cylinder and its connections, and suppose the cylinder is placed at zero potential by connecting it to earth. Insulate and note the reading after some suitable interval of time, say t seconds. .Let V be the potential corresponding to this reading; this can easily be determined by calibrating the instrument with suitable known potentials. The charge given to the electrode in the time t is then CV. so that the current between P and F is CV/t.

Since the capacity of a system can be made very small the method is far more sensitive than the most delicate galvanometer. The capacity of an electroscope of the C. T. R. Wilson type with its electrode and all its connections need not be greater than 4 or 5 cms.; or, say, 5×10^{-12} farads. The electroscope will easily give a deflection of 100 divisions for a potential of 1 volt, while a movement of the gold leaf of a division per minute can be detected with certainty. This corresponds to a current of 10^{-15} amperes. The best galvanometers give a perceptible deflection for a current of about 10^{-11} amperes. The electroscope method is, therefore, about ten thousand times as sensitive as the most delicate galvanometer.

In the present case the value of $C.V./t$ gives the charge carried by the α-particles per second from the upper surface of the polonium disc. In electrostatic units it was found to be ·000377. This was the charge carried by the $3·935 \times 10^5$ α-particles emitted by the polonium. The charge on each particle was therefore $·000377/3·935 \times 10^5$ or $9·58 \times 10^{-10}$ in electrostatic units. Rutherford's value was $9·30 \times 10^{-10}$ in the same units.

It will be seen that these numbers are almost exactly twice the value $4·7 \times 10^{-10}$ *e.s.u.*, which represents the mean value for the electronic charge e.

In fact the value of $2 . e$ is almost midway between these experimental numbers. The charge on the α-particle is therefore exactly twice the charge on the electron. The α-particle is therefore an atom which has lost two electrons. This is characteristic of all the α-particles emitted by radioactive substances which appear to differ from each other only in velocity.

We have stated that the α-particle is an atom; the question arises of what element is it an atom? The ratio of the mass to the charge can be determined for the α-particle in the same way as for the electron, neglecting the different experimental arrangements required. No great accuracy has, however, been reached, as the deflection of these particles from radioactive substances even in the strongest magnetic fields is small. According to Rutherford, the ratio of mass to charge for the α-particle from radium is $1{\cdot}97 \times 10^{-4}$, and we have seen that they carry two charges, or in the language of electrolysis, are divalent. The electrochemical equivalent of hydrogen is 10^{-4}, so that if we can assume that the ion in solution carries the electronic charge e, the atomic weight of the α-particles would be 3·84, which is very near 3·96, the atomic weight of helium. Conversely, if we can show that the α-particle is an atom of helium we shall have direct evidence that our assumption that the monovalent ion carries the same charge as an electron is correct.

The proof of the nature of the α-particle has been given by Rutherford to whom we owe so much of our knowledge of these rays. The experiment is based on the fact that the α-rays, owing to the great speed with which they are projected by the parent molecule, are able to pass through thin sheets of solid substances which are perfectly impervious to the molecules of ordinary gases.

The substance used as the source of α-particles

was radium emanation, the radioactive gas which is the first decomposition product of radium, and on which Professor Ramsay proposes to bestow the name of niton. This gas, which is disengaged from solutions of radium in water, after being purified in various ways, is finally passed into a fine thin-walled glass tube, the end of which has been carefully sealed. This is enclosed by a wider thick-walled glass tube which is also made quite air-tight and has in its upper portion two electrodes. The spectrum of the discharge between these two electrodes enables us to determine the nature of the gases in this outer tube. There is no connection between the inner and outer tubes, so that nothing can pass from the one to the other, except the α-rays. This point was tested by filling the inner tube with helium and allowing the apparatus to stand for some days. No trace of the helium spectrum was obtained in the outer tube under these circumstances. When, however, the inner tube was filled with emanation, the α-particles emitted were able to pass through the thin glass walls and, being stopped by the thicker tube beyond, collected in the space between the two tubes.

Slowly, as the experiment went on, the spectrum of helium began to appear, becoming brighter and more complete as more and more of the rays passed through the inner tube, until at the end of a few hours the complete helium spectrum was obtained. The blank experiments were far more in number and severity than has been described here. They left no doubt that the α-particles were indeed atoms of helium, differing from the ordinary atoms of the gas only in their velocity and charge. We have seen above that this result leads directly to the conclusion that the charge on a monovalent ion in solution is the atom of electricity which we denote by e.

This is one of the most direct proofs of this very

important fact. There are, however, many others of very diverse natures, some of which we shall have occasion to discuss when dealing with the phenomena on which they are founded.

The charge carried by a monovalent ion in solution and the electronic charge are thus the same; the value for the latter is, as we have seen, very near $4{\cdot}7 \times 10^{-10}$ *e.s.u.*, or $1{\cdot}57 \times 10^{-20}$ in our electromagnetic units. The ratio of the mass to the charge for the hydrogen atom is, as we have seen, $1{\cdot}04 \times 10^{-4}$ Thus the mass of the hydrogen atom is $1{\cdot}04 \times 10^{-4}\,e$, or $1{\cdot}63 \times 10^{-24}$ **gms.** This is therefore the value in grammes of the chemical unit of atomic weight, and we can at once calculate from it the actual mass of any atom of which the chemical atomic weight is known. The accuracy of the result depends only on the accuracy with which the value of *e* can be determined. A reference to Table II. (see page 27) will show that the error in the value we have used probably does not exceed 2%, or 3% at most.

Since the mass of a cubic centimetre of hydrogen at normal temperature and pressure is almost exactly 9×10^{-5} gms., the number of hydrogen molecules contained in 1 cub. cm. of a gas under these conditions is $9 \times 10^{-5}/2 \times 1{\cdot}58 \times 10^{-24}$, or $2{\cdot}75 \times 10^{19}$. By Avogadro's hypothesis this number is independent of the nature of the gas.

CHAPTER IV.

THE NEW METHOD OF ANALYSIS.

We have considered so far the case of the positively charged particles from radioactive substances, and have seen how their mass, charge, and chemical nature have been determined, and how these determinations have given us the most certain and accurate estimate of the actual mass of a hydrogen atom. The information obtainable was, however, limited by the fact that these particles are all of one kind, consisting always of atoms of helium.

We have already mentioned that it is possible, in a discharge tube, to endow atoms and molecules of ordinary matter with those properties of high velocity and electric charge which have made the *a*-particles so amenable to experiment. The process may be regarded as follows. When the two electrodes of the discharge tube are connected to the poles of an induction coil, the molecules of the rarefied gas in the tube are subjected to a very strong electric tension. This tension acting upon the negative electrons, which, as we have seen, are contained in all molecules, causes one or more of them to be displaced from the molecule, leaving the remainder positively charged. This positive particle, as we may now call it, is driven with great velocity towards the cathode by the action of the electric field which produced it.

If the cathode is suitably perforated, the positive particle will pass through it, and emerge on the further side with this velocity unimpaired. It can then be detected by allowing it to fall either on a fluorescent screen, or preferably on a photographic plate, where it will produce a permanent record, which can be measured at leisure with suitable instruments of precision.

The experiments, for reasons to which we have already alluded, presented many difficulties. It

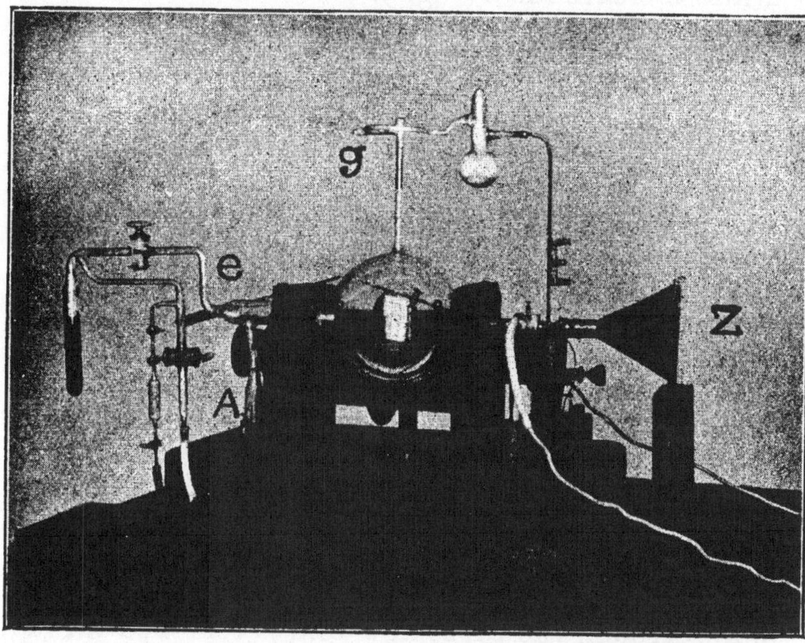

FIG. 8.—THOMSON'S POSITIVE RAY APPARATUS.

is only after several years of experiment that Professor Sir J. J. Thomson has at last perfected an apparatus which allows him to measure these particles with accuracy and certainty.

A general idea of the apparatus can be obtained from Fig. 8, which is reproduced from a photograph of one of Professor Thomson's tubes. The details are shown more clearly in the accompanying diagram, Fig. 9, which represents a section of part of the apparatus as seen from above.

The apparatus consists of two parts: the discharge tube proper, where the particles are produced and the "camera," where their presence is detected and their masses determined. The discharge tube proper consists of a large bulb T, some 30 cms. in diameter. The anode A (Fig. 8) is placed in a side tube, while the cathode C (Fig. 9) occupies the neck of the bulb. The gas to be experimented on enters the

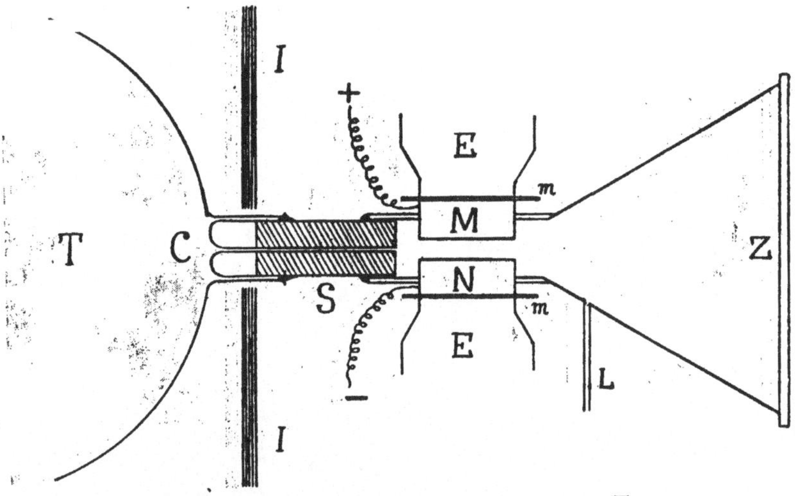

FIG. 9.—THOMSON'S POSITIVE RAY TUBE.

bulb through the tubes at *e*, and is removed through the tube *g* by a Gæde pump, which is worked sufficiently rapidly to keep the gas in T at a suitably low pressure.

The cathode C (Fig. 9) is an aluminium rod pierced with a very fine copper tube about 8 cms. long and less than 1-10th mm. in diameter. An exceedingly fine pencil of positive rays is thus obtained. It is evident that with a tube so narrow any stray magnetic field will cause the particles to hit the walls of the tube and be lost. To prevent this the copper tube is enclosed for the greater part of its length in a thick iron tube S, shaded in the diagram, while thick iron screens I, I, protect the main discharge from the influence of the magnet.

Owing to the intensity of the discharge, the cathode becomes very hot, and it is necessary, in order to protect the numerous sealing-wax joints, to surround them with a water cooling jacket, as shown in Fig. 8.

On leaving the copper tube, the rays pass between M and N, two soft iron blocks let in through the sides of the tube (which is here made of ebonite) and carefully waxed round, so as to be completely vacuum tight. These blocks serve two purposes. In the first place, they form the pole-pieces of the electromagnet E, E, from which, however, they are electrically insulated by two thin strips of mica *m*, *m*. This insulation enables them to be charged positively and negatively respectively, by means of a cabinet of small storage cells, and thus to serve as electrodes for the application of an electrostatic field.

In this way the positive particles can be subjected at the same time to a magnetic and an electric field. It will be noticed that, whereas in the case of the cathode ray experiments described in an earlier chapter the two fields were at right angles to each other, in this case the two fields are parallel, and the corresponding deflections therefore at right angles to each other.

After passing through these fields, the rays fall either upon the glass plate Z, which is covered with powdered willemite, or upon a photographic plate, which can be introduced when required immediately in front of Z by suitable devices.

The side tube L leads to a tube filled with charcoal and immersed during the experiment in liquid air, a device due originally to Sir James Dewar for absorbing as completely as possible the gases in this part of the apparatus. This detail, apparently insignificant in itself, has had much to do with the success of the method. If the gas in the camera is at the pressure most suitable for the production of positive rays in the discharge tube, the effects observed on the screen are found to be almost entirely secondary. The residual gas in the camera becomes strongly ionized by the action of the positive rays upon it. The latter are thus travelling through an atmosphere charged with negative electrons, with which they readily combine to form neutral molecules again. When this has taken place, the particle is no longer acted upon by electric or magnetic fields. The exact extent of its deflection will depend upon the length of time during which it succeeded in retaining its charge. These secondary effects still make their appearance upon some of the photographs taken. They can, however, always be detected by the fact that they either disappear or alter in position or appearance, if the pressure in the camera is slightly lowered or slightly raised.

The charcoal and liquid air method enables us to remove the residual gas so rapidly that it is possible to have the pressure in the camera considerably lower than that in the discharge tube, the diffusion along the long narrow copper tube, which is their only connection, being comparatively slow. In this way the production and the measurements of the particles can both be carried on under the most favourable conditions.

Let us consider the course of the positive particle from its genesis in the discharge tube to its stoppage on the photographic plate. Particles formed near the anode will fall through practically the whole of the difference of potential between the electrodes of the discharge tube, some 30,000 or 40,000 volts. The energy they acquire will thus be V × E where V is this difference of potential, and E the charge on the particle. Their velocity v, on reaching the cathode, will thus be given by

$$v^2 = 2.\mathrm{VE}/m.$$

This is the maximum velocity the particles can

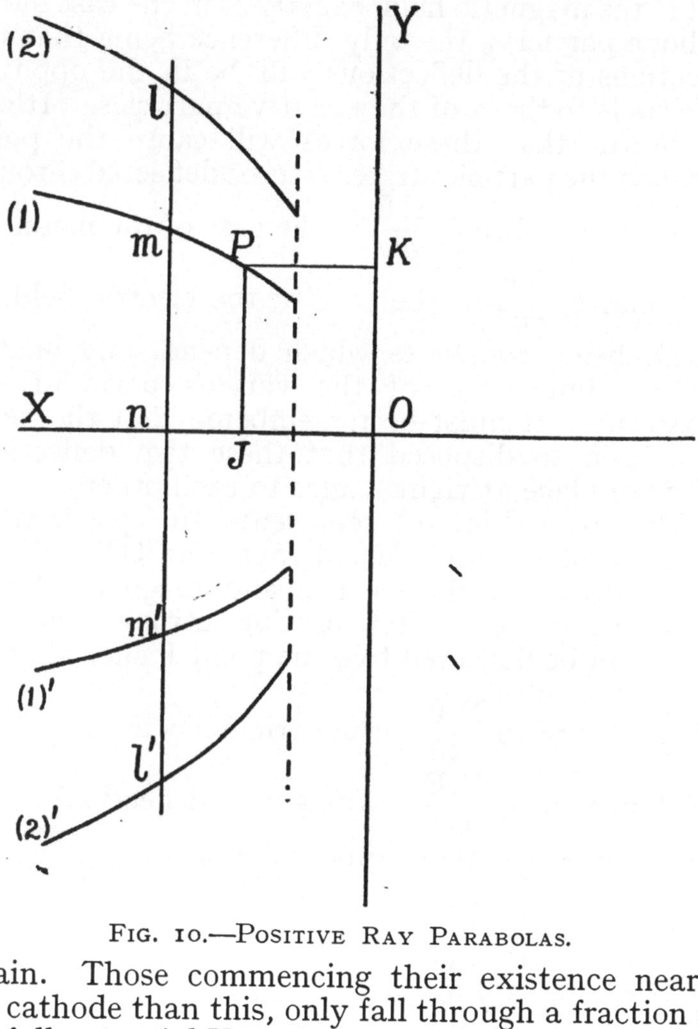

FIG. 10.—POSITIVE RAY PARABOLAS.

attain. Those commencing their existence nearer the cathode than this, only fall through a fraction of the full potential V, and thus acquire only a fraction of this maximum velocity. Thus, the stream of rays passing through the cathode will contain particles moving with all velocities up to a certain maximum.

On passing into the electric and magnetic fields at MN they will be acted upon by forces equal to XE and H Ev respectively where X is the electric and H the magnetic field, exactly as in the case of the cathode particles, the only difference being that the directions of the deflections will be in the opposite directions to those of the negative particles. It can be shown* that these forces will cause the point at which the particle strikes Z to be deflected through distances equal to $k_1 \frac{\mathrm{HE}}{mv}$ in the case of the magnetic field, and $k_2 \frac{\mathrm{XE}}{mv^2}$ in the case of the electric field, k_1 and k_2 being constants which depend only on the relative dimensions of the various parts of the apparatus. It must be remembered that the fields have been so disposed that these two deflections will take place at right angles to each other.

Thus if O (Fig. 10) represents the position on the plate of the undeflected rays, and OX, OY are the directions of the electric and magnetic deflections, the particle after passing through the two fields, will be deflected to some point P such that

$$\mathrm{PK} = x = k_2 \frac{\mathrm{X.E}}{mv^2} \quad \text{(electric deflection)} \quad . \quad \text{(a)}$$

$$\mathrm{PJ} = y = k_1 \frac{\mathrm{H.E}}{mv} \quad \text{(magnetic deflection)} \quad . \quad \text{(b)}$$

We see from these equations that

$$y/x = \left(\frac{k_1}{k_2}\frac{\mathrm{H}}{\mathrm{X}}\right) . v \quad . \quad . \quad . \quad . \quad . \quad . \quad . \quad \text{(c)}$$

$$y^2/x = \left(\frac{k_1^2}{k_2}\frac{\mathrm{H}^2}{x}\right)\frac{m}{\mathrm{E}} \quad . \quad . \quad . \quad . \quad . \quad . \quad \text{(d)}$$

Thus, if the strengths of the two fields are kept

* Appendix A.

the same throughout the experiment, all the quantities within the brackets will be constant. Thus, y/x will be a measure of the velocity of the deflected particle, while the ratio y^2/x will measure the ratio of the mass to the charge.

If all the particles were identical, and had the same velocity, the spot O would simply be shifted to the point P. As we have seen the velocities of the particles vary considerably. The rays will therefore not be deflected to a single spot, even if their mass and charge are identical, but will be drawn out into a line or band. By equation (d) all the particles for which m/E is the same will, however, lie on some curve for which the value of y^2/x is constant; such a curve is known as a parabola.

If the particles emerging from the cathode consist of atoms of differing atomic weight, they will be sorted out into a series of parabolas one above the other, each curve corresponding to some definite value of m/E. Since the velocity which is proportional to y/x cannot exceed a certain value, these curves will not be complete, but will stop short at some definite distance from the axis OY. The appearance of the plate after exposure should therefore be somewhat as shown in the top left hand quadrant Fig. 10, where two such parabolas are shown.

To measure the curves, drop any perpendicular on OX, cutting the curves in l and m, and the axis in n. Then, from equation (b) it can be seen that

$$ln^2/mn^2 = y_2^2/y_1^2 = \left(\frac{m_2}{\mathrm{E}_2}\right) \div \left(\frac{m_1}{\mathrm{E}_1}\right)$$

since the value of x is the same for all points along the line lmn. Thus, if $\mathrm{E}_1 = \mathrm{E}_2$, that is to say, if the two particles carry the same charge, the ratio of the masses of the two sets of particles is simply equal to ln^2/mn^2.

The line OX has no real existence on the plate, but the apparatus can be arranged so that the direction of the deflections are parallel to the edges of the photographic plate, while the position of O is always marked by a very bright spot caused by rays which come through the cathode uncharged, and are therefore undeflected. A more

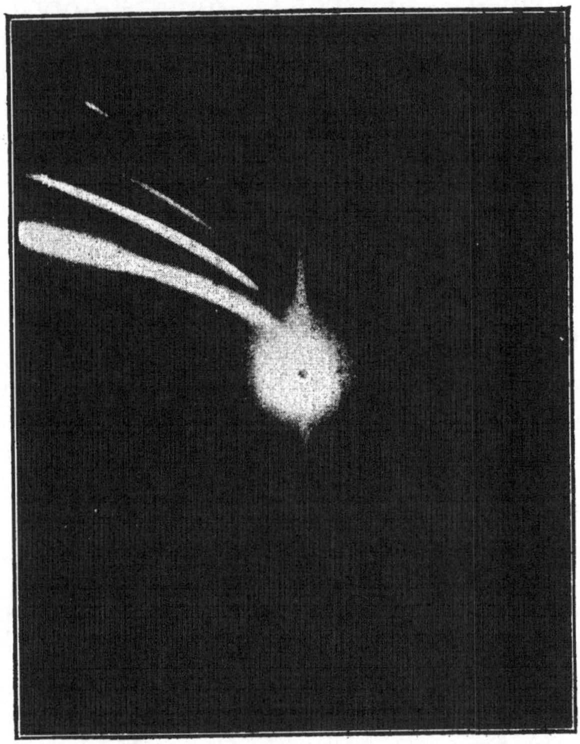

FIG. 11.—PARABOLAS DUE TO MULTIPLE-CHARGED MERCURY ATOMS.

accurate method is to reverse the field halfway through the exposure. The direction of the deflection is thus reversed, and symmetrical parabolas are formed on the opposite side of OX as shown in Fig. 10. The distance between the two branches of the same curve is then twice the distance of either from the line OX.

It will have been noted that the ratio of the squares on these ordinates does not give directly the ratio of the masses of the particles, but only the ratio of what we may term their electrochemical equivalents. Fortunately, however the majority

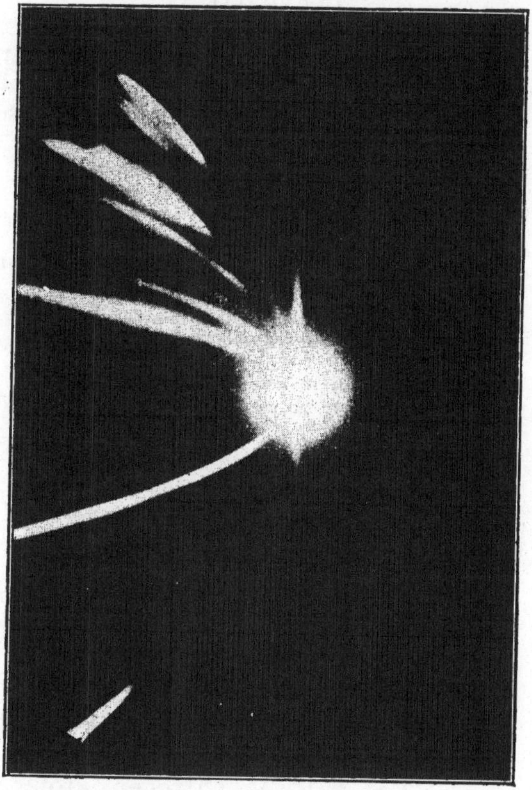

FIG. 12.—TYPICAL POSITIVE RAY PHOTOGRAPH.

of the particles carry the single electronic charge *e*, while in any case the charge E on the positive particle cannot be more than a small integral multiple of this. Very little uncertainty is as a rule introduced from this cause, though it does occasionally arise. Mercury is the greatest offender in this respect, as mercury atoms carrying as many as eight charges are sometimes to be detected. Thus, five of the six parabolas on Fig. 11, which is reproduced from one of Prof. Thomson's photographs, are due to atoms of mercury, carrying respectively one, two,

TABLE III.

Atmospheric nitrogen. Potential across discharge tube, 30,000 volts; current through magnet, 3·5 amperes; potential difference in electric field, 200 volts.

d m.m.	*m*/E.	Probable cause of line.	
7·2	200	Hg +	Mercury atom with single charge.
10·3	100	Hg++	Mercury atom with two charges.
12·4	67	Hg +++	Mercury atom with three charges.
15·4	44	CO_2 +	Molecule of carbon dioxide.
16·5	39	A +	Argon atom (40) with single charge.
19·4	28	N_2 +	Nitrogen molecule with single charge.
23·1	20	Ne +	? Neon with single charge.
25·6	15·9	O +	Atom of oxygen with single charge.
27·6	14	N +	Nitrogen atom with single charge.
30·0	12	C +	Carbon atom with single charge.
38·7	7	N ++	Nitrogen atom with two charges.

three, four, and eight charges. The latter line, though clear on the negative, is too faint to show up clearly in the reproduction.

The measurement and interpretation of the results can best be illustrated by an actual example.

Fig. 12 represents a fairly typical photograph obtained with the apparatus. The gas in the discharge tube in this case was atmospheric nitrogen, that is to say, air from which the oxygen had been extracted. Table III. contains the actual measurements made on this plate.

The first column gives the distance of the different parabolas from the axis OX, measured along some common ordinate as explained above. The second column gives the value deduced from these figures for the relative masses of the particles, assuming that they all carry the same charge. If the charge carried is double or treble this value, the corresponding mass so obtained will be half, one-third and so on, of the actual mass of the particle. The standard line on this plate was the innermost one of all, which as its appearance shows, is due to mercury carrying a single charge. The mass of this particle is for convenience taken as being equal to its atomic weight (200).

Examining the table, it will be seen that the three inner parabolas represent particles having a value of m/E of 200, $\frac{1}{2}$. (200), and $\frac{1}{3}$ (200). They are due to mercury atoms having, respectively, one, two, and three charges. Mercury vapour is always present in the tube unless special precautions are taken to exclude it, as it is used in the vacuum pumps attached to the apparatus. The next line is made by particles having a mass of 44 ; these are singly charged molecules of carbon dioxide ; carbon compounds being usually present as impurities in small quantities.

The argon line (39) is the long bright parabola, immediately below the very thick bright line (28), due to molecules of nitrogen with a single positive charge. This, as might be expected, is the brightest line in the photograph. The parabola for which d is 23 . 1, is very faint on the reproduction. It probably represents Neon (20) though there is also the possibility that it is formed by argon atoms with a double charge, as the argon line (40) is very bright.

FIG. 13.—POSITIVE RAY PARABOLAS.

Our method does not enable us to distinguish between these alternatives.

The triplet of lines which follow is due to atoms of oxygen, nitrogen, and carbon. The last line on the plate, indicating a value 7 is the nitrogen atom with a double charge. The hydrogen line which is nearly always present, has been deflected entirely off the plate by the large magnetic field applied in order to separate the heavier constituents.

It will thus be seen that all the lines on our plate correspond to elements which we might expect to find present from the nature of the gas employed. There are no lines unexplained, and the values obtained for the relative masses of the particles correspond very closely to the atomic and molecular weights determined from chemical considerations.

The faint lines seen on Fig. 12 joining some of the parabolas to the origin, are due to the secondary effects which we have already mentioned. The mercury parabolas, however, as will be seen from the figure, do actually extend to within a very short distance of the origin. This is due to quite a different cause. Turning back to equation (a), it will be seen that the electric deflection is proportional to the charge E_1 carried by the particle in the electric field divided by $\frac{1}{2}mv^2$, the kinetic energy of the particle. We have seen that the maximum value of the latter is equal to $V . E_2$, where V is the potential of the discharge, and E_2 the charge carried by the particle in the discharge tube. The minimum value of the electric deflection, that is to say, the nearest point to which the parabolas can extend will be proportional to E_1/E_2 and will thus be the same for all particles, whatever their mass and charge, if each particle carries the same charge in the tube as in the camera. Thus all the curves will stop short at a certain line drawn parallel to OY. This is well shown in Figs. 13 and 14.

If, however, a particle in the discharge tube having a double or treble charge, loses one or more of these extra charges before the electric field at MN is reached, the value of E_2/E_1 will be greater than unity, and the parabola will extend further towards the axis OY.

These extensions of the parabolas are very clearly shown in the three inner curves of Fig. 13, where an almost complete break occurs between the termination of the original parabola and its continuation by particles which have parted with one of their original charges. These three parabolas are due respectively to Mercury (200), Xenon (132), and Krypton (83). The two outer curves, representing

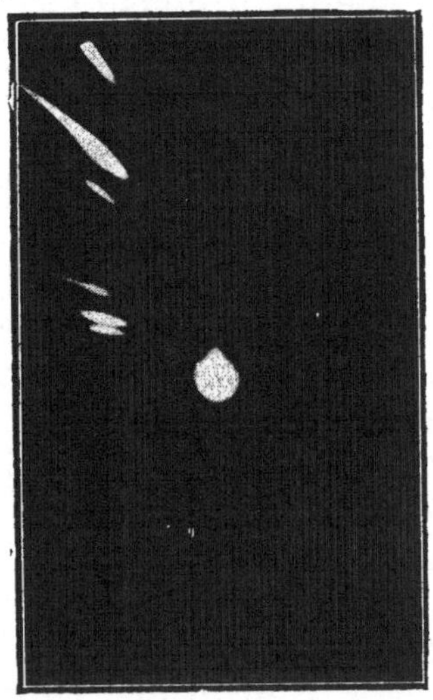

FIG. 14.—GAS EVOLVED FROM PLATINUM.

Argon (40) and the Nitrogen molecule (28), end abruptly at the normal distance from OY.

The best example of the effect is, however, shown by the mercury lines in Fig. 11. The normal mercury atom in the discharge tube apparently carries a charge of $8e$, and approaches the cathode with an amount of energy equivalent to this. The innermost mercury parabola (Fig. 11) gives a value for the mass just 200 times that of the hydrogen atom. It is therefore due to a mercury atom carrying a unit charge, and having thus only one-eighth of the charge it carried in the discharge tube. The value of E_1/E_2 is thus $\frac{1}{8}$ and the parabola extends to one-eighth of the usual minimum distance from the vertical axis. The next line ($m = 100$) is due to atoms with two charges, and extends to one-quarter of the usual minimum distance, and so on, until the last parabola which corresponds to a mercury atom with eight charges (the same number as in the discharge), ends up at the usual distance from the axis.

This gradual shortening of the parabolas for the differently charged mercury atoms, is beautifully shown in Fig. 11, where the terminations of the five mercury parabolas can be seen to lie exactly on a straight line through the point O inclined at an angle to OY, as would be expected on the above theory.

It will be noticed that the lines decrease in intensity as the number of charges carried is increased. This is always to be observed on the photographs, and is a fact which sometimes helps us to the right interpretation of a doubtful line in a photograph. For example, if the line indicating a mass of 20 on Fig. 12 had been brighter than the Argon line (40) we should have been able to say definitely that it was due to atoms of Neon, and not to Argon atoms with a double charge.

The plates so far described were all taken using large magnetic fields in order to obtain a suitable dispersion of the heavier atoms. The last plate of the series (Fig. 14) was taken with a smaller magnetic dispersion, and shows the lines due to some of the lighter elements. It is of peculiar interest at the present moment, as it is one of the photographs which shows the mysterious line due to particles for which the ratio of the mass to the charge is exactly three times that for the singly charged hydrogen atom. On this account it may perhaps be of interest to give the actual measurements made on this plate (Table IV.). The first column of the table gives the actual measurements of a common ordinate of the different parabolas, the second column the corresponding values of m/E, the ratio for the hydrogen atom being taken as unity.

The gas used in the experiment was a sample of that mysterious gas which is evolved in considerable quantities when any metal is heated, or better still, as in the present case, bombarded with cathode rays in a vacuum tube. That gases were evolved under these circumstances had been known for some considerable time, but their analysis had not been

TABLE IV.

Gases evolved from Platinum bombarded with Cathode rays for 6 hours. (See Fig. 14.)

d m.m.	*m*/E.	Probable cause of line.
27·0	1·00	H + Hydrogen atom.
19·1	2·0	H_2 + Hydrogen molecule.
15·6	3·0	? Unknown element.
7·8	12	C + Carbon atom.
6·7	16	O + Oxygen atom.
5·8	21	Ne + Neon atom (20).
4·8	32	O_2 + Oxygen molecule.

previously effected. As will be seen from the table, the gas consists principally of hydrogen, but helium and neon are also present; while the plate also shows traces of carbon and oxygen, probably due to impurities in the apparatus.

In addition to these known substances there is also very clearly marked a line, the third from the top of the plate, which cannot be assigned to any known element or compound. There was at first the possibility that it might be due to a carbon atom carrying four elementary charges, but this hypothesis was found to be untenable. In the first place the line under consideration is much brighter than that due to the singly charged carbon atoms, while we have seen that on a given plate the curves due to multiply charged atoms are invariably fainter than those due to atoms of the same substance having a single charge. In the second place, when precautions were taken to exclude carbon compounds from the apparatus the carbon line (12) completely disappeared, while the line 3 remained as bright as ever.

Only two possibilities remain. The line may be due to a previously unknown element of atomic weight 3, or it is due to a hitherto unknown modification of hydrogen H_3 corresponding to ozone. If the latter is the case, the substance has some quite unexpected properties, as it is certainly not destroyed by sparking with excess of oxygen. Professor Sir J. J. Thomson is still actively pursuing his experiments on this substance, and some very interesting results may be expected. Until these experiments are completed it would be premature to venture any further opinion as to the real nature of this new gas.

Another parabola which is found when the gases from liquid air residues are introduced into the tube suggests interesting possibilities. This is a line giving a value for m/E equal to 22. This might possibly be a molecule of carbon dioxide (44) with a double charge; but from the fact that it appears only in these mixtures of the rarer atmospheric gases, and that it is quite well marked even when the CO_2 and CO lines are absent seems to point to it being a new compound NeH_2 of neon and hydrogen.

A few further points of interest in the photographs may be noted. The bright central spot shown in all the figures is due to the particles which have lost their charge during their passage through the long copper canal tube, and are therefore undeflected by any combination of electric and magnetic fields. The atoms of some elements have such a strong attraction for the surrounding negative electrons that they actually take up more of them than would be sufficient to neutralise the original positive charge. They thus enter the deflecting fields with a negative instead of a positive charge, and so give rise to parabolas in the opposite quadrant to those due to the positive particles, but giving, of course, the same values for m/E. At least two of these "negative" parabolas appear on Fig. 13. Oxygen and the halogens if present in any quantity almost invariably give these negative curves, Strangely enough hydrogen, which is generally reckoned among the more electro-positive elements, also frequently shows the same effect. It has never been found with nitrogen or helium.

We have so far only been considering the relative values on the ratio m/E for the different particles. By measuring the dimensions of the different parts of the apparatus and calculating the actual values of the constants k_1 and k_2, the absolute value of this ratio for any given parabola can be deduced. Such a determination was made by Professor Sir J. J. Thomson, and it was found that the absolute value for the outermost parabola of all was almost exactly 10^{-4}. We have seen that this is the value of the ratio for a hydrogen ion in solution or a hydrogen atom carrying a single charge. The particles with which we have been dealing are therefore actual atoms and molecules of the substances in the discharge tube, as we have tacitly assumed in the foregoing pages.

We have now described in some detail the principal features of the new method of molecular analysis. Its importance can hardly be exaggerated. In the first place it enables us to detect with certainty the presence in a gaseous mixture of far smaller quantities of an element than could be accomplished by any other method known to science. Professor Thomson has given the following striking illustration of the delicacy of the method. It was found that the photographs taken with atmospheric argon in the apparatus always showed the helium line faint but quite distinct. In this particular experiment the volume of the discharge tube was about 2 litres, and the pressure of the gas in the tube 1/300 m.m. of mercury. The gas experimented on would therefore have had a volume of about 1/100 c.c. at atmospheric pressure. This is approximately the quantity of argon in 1 c.c. of air at atmospheric pressure. We are thus able to detect the helium present in a single cubic centimetre of air. According to the determinations of Professor Ramsay this amounts to no more than four millionths of a cubic centimetre.

The new method has the further advantage over spectroscopic analysis that it gives directly the atomic or molecular weight of the substance under experiment. Thus the photograph of Fig. 14 not only reveals the existence of a new gas but gives at once its molecular weight.

Lastly, and this is perhaps most important of all from the point of view of an intimate study of the modes and methods of chemical combination and decomposition, the time taken for each single particle to register its presence and mass is exceedingly minute. A very moderate value for the speed of the positive particles is 10^8 cms. per second (about 620 miles per second). Their path in the discharge tube is not more than 50 cms. in all. Thus a particle when caught by the discharge will register itself on the photographic plate in considerably less than one millionth of a second.

We might therefore reasonably expect to find on the photographs curves corresponding to temporary combinations of atoms so unstable that their whole period of existence is not greater than one millionth of a second, if any such combinations are formed in the discharge tube. Thus in the case of chemical reactions taking place within the discharge tube we might expect to find on the plate traces not merely of the stable products of the reaction but also of any intermediate stages occurring in the course of the reaction. The field thus opened up for an intimate study of the mechanism of chemical reactions is illimitable.

Though little has as yet been accomplished along these lines, the experiments already made show that this expectation is not unfounded. As a very simple illustration of the application of the method we may take the case of phosgene, $COCl_2$. When this was introduced into the apparatus the photographs showed that, in addition to the molecules of undecomposed phosgene (99), the carriers of positive electricity were molecules of carbon monoxide (28) and atoms of chlorine (35·5). The lines due to single atoms of carbon and oxygen were very faint indeed, thus showing that the decomposition of the compound by the discharge consisted of the separation of the chlorine atoms from the CO molecule, the bond between the carbon and oxygen atoms remaining intact.

This is a simple case where the results obtained might easily have been predicted. Another interesting case is that of methane, CH_4. When this gas is subjected to the discharge a group of five lines appears on the photographic plate, indicating the presence of particles having masses 12, 13, 14, 15 and 16 times that of the hydrogen atom. These correspond to particles having the composition C, CH, CH_2, CH_3, and CH_4 respectively.

We have been dealing so far with the many advantages of the method. Some of its limitations should perhaps be pointed out. In the first place the substance to be exeprimented on must be either gaseous or capable of exerting a vapour pressure at ordinary temperatures. A more important defect, and one which is shared by the spectroscopic method, is that an element may be present in considerable quantity in the discharge tube without producing a corresponding trace upon the photographic plate. Thus, although many metals have compounds sufficiently volatile for the purpose of the experiment, mercury and nickel are the only metallic elements which have so far been detected by the method.

One word may be added as to the relative intensity of the different lines on a given photograph.

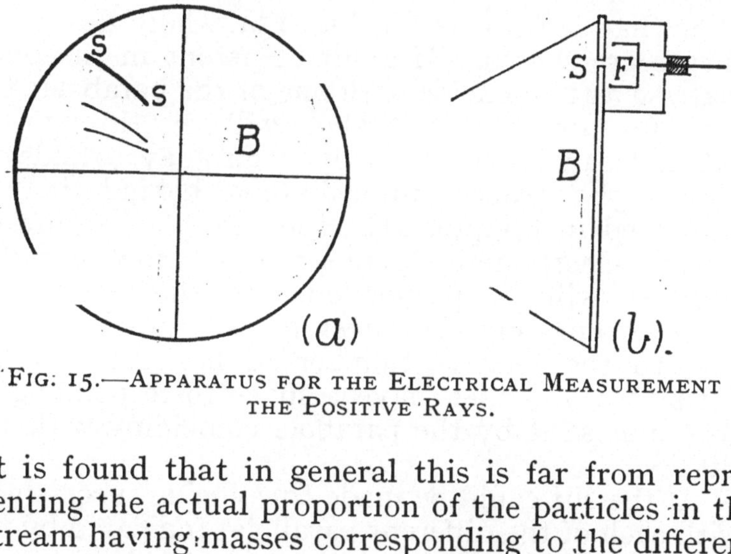

FIG. 15.—APPARATUS FOR THE ELECTRICAL MEASUREMENT OF THE POSITIVE RAYS.

It is found that in general this is far from representing the actual proportion of the particles in the stream having masses corresponding to the different lines. For some reason at present unexplained, a given number of atoms of a light element produce far greater effect on a photographic plate or willemite screen than an equal number of the atoms of a heavier element. Thus the parabola corresponding to the hydrogen atom will often be the most intense line on the plate when the amount of that element present forms less than 1% of the contents of the tube.

This has been very clearly demonstrated by Sir J. J. Thomson by substituting an electrical for a photographic method of detecting the positive particles. It must be remembered that as all the particles with which we are dealing carry an electric charge they can be measured by collecting them in a Faraday cylinder in the way described for the α-particles in the preceding chapter.

To carry out the experiment a brass plate B (Fig. 15) was substituted for the glass screen Z of Fig. 9. If the brass plate is covered with powdered willemite a series of bright parabolic curves will of course appear upon it when the tube is running somewhat as indicated in Fig. 15a, where three such curves are shown. If a slit SS is cut in the brass plate so as to coincide with one of the parabolas the particles forming this parabola will pass through the slit and can be collected by a Faraday cylinder F (Fig. 15b) placed immediately behind it. As explained in Chapter III., the rate of charging up of an electrometer connected to the Faraday cylinder will be proportional to the number of particles entering the latter per second, that is to say, to the relative number of molecules in the positive rays which have a mass corresponding to that indicated by the parabola coinciding with the slit SS.

If the slit could be made to coincide successively with each of the different parabolas indicated by the glowing willemite, the relative rates of deflection of the electrometer for the different positions would be proportional to the number of particles of corresponding mass in the positive rays from the discharge. The experiment can be performed much more conveniently without moving the slit by varying the intensity of the deflecting magnetic field, H. If this is increased it can easily be seen that the corresponding magnetic deflections of the particles will also be increased and the various parabolas will move outwards from the centre of the screen. In this way each of them in turn can be brought to coincide with the parabolic slit SS. It can easily be seen from equations (a) and (b) that if the other conditions of the experiment remain the same the value of m/E for a given set of particles is proportional to the square of the strength of the magnetic field necessary to deflect them to a given position. Thus from the various rates of deflection of the electrometer corresponding to different values of the magnetic field we can at once determine the relative number of particles in the beam of positive rays corresponding to definite values of the ratio m/E. In other words, we can determine the relative proportions of the different kinds of atoms and molecules which make up a given beam of positive rays.

On making the experiment it was found that there was practically no movement of the needle of the electrometer connected to the Faraday cylinder except when one of the bright parabolas was made to coincide with the slit. If a fairly narrow slit was used, the appearance and disappearance of a deflection of the electrometer as such a line passed across the slit was surprisingly sharp, and lines quite close together could easily be separated.

Although we can thus perform a quantitative analysis of the beam of positive rays it cannot be assumed without further experiment that the number of atoms of an element in the beam is necessarily proportional to the quantity of the substance present in the discharge tube. Preliminary experiments, however, seem to indicate that this is very approximately the case at any rate for the non-metallic elements. If so, this latest modification of the method of molecular analysis will enable us to carry out not merely a qualitative but also a quantitative analysis of any gas introduced into the discharge tube. The quantitative determination of traces of a gaseous substance too small to be detected by the spectroscope will then have become an accomplished fact.

CHAPTER V.

THE NATURE AND SIZE OF AN ELECTRON.

WE must now turn from these triumphs of experimental skill to some theoretical considerations, and discuss briefly the new views of the atom to which they have given rise. It would be idle to pretend that we have as yet any theory of the structure of the atom which absolutely and completely sums up all the known phenomena of Science. The scientific philosophers of an earlier age hoped that at no very distant date they might be able to include in one or more ponderous tomes all that could be known in the realm of physical phenomena. Looking down the long vistas which the new discoveries have so recently opened up, we expect no such immediate end. We are not yet within sight of a mechanical or electrical theory of the universe wherein all the chemical and physical properties of the material universe shall be explicable in terms of the electrons and their motions. Still, while we cannot say that we have such a theory, there are certain sign-posts pointing out the direction in which we may hope for ultimate success. Even now it is possible to construct "models" of the atom which, however crude and insufficient they may be, enable us to link together many most diverse phenomena.

In the realm of science an insufficient, or even a false, hypothesis is better than none at all, and it will be useful to take the best approximation we can devise to serve us as a clue through the labyrinth of experimental facts which we have still to consider. At the same time, it will be well to point out carefully where we are treading the solid ground of experimental fact and legitimate deduction, and where we rise to the higher but more precarious flights of scientific imagination.

We have seen that electrons are contained in all atoms, and many phenomena lead us to believe that they play a most important part in determining both the chemical and the physical properties of the substance. The question naturally arises, what is an electron? Fortunately to this question also we are able to give some answer. An electron is an atom of electricity free from association with anything in the nature of matter as we know it. It not merely has a charge, it is a charge, and apart from its charge has no existence and no properties, not even that of mass.

The exact proof of this statement has involved long and intricate mathematical analysis. It will, however, be possible, without going at all deeply into these abstruse calculations, to make clear the nature of the proofs advanced, and the experimental evidence upon which they rest.

In the first place, since we have already determined the mass of an electron, we must show that such a moving electric charge will possess mass, a property which we have been accustomed to associate only with what we call matter. The proof follows directly from the ordinary laws of electric and magnetic induction, and was first given by Professor J. J. Thomson, in 1881, many years before the discovery of the electron rendered possible any practical application of the result.

Consider a small sphere charged with electricity of either sign moving through space with a velocity v. We have already pointed out that this moving sphere will be equivalent to an electric current element coinciding with the path of the particle. By the usual laws of electro-magnetic induction this current element will produce a magnetic field, at a distance r from the particle, equal to $ev \sin \theta / r^2$, where θ is the angle between the direction of motion of the particle and the direction of r. Thus, the moving particle is surrounded not only by an electro-static field due to its charge, but also by a magnetic field due to the motion of that charge. It will be seen that this field has its greatest value in a direction at right angles to the motion of the particle, and falls to zero in the direction in which the particle is travelling.

It is a well-known fact that a magnetic field is the seat of energy, the energy in a field of strength H being $\frac{H^2}{8\pi}$ times the volume. The magnetic energy in a small space of volume u at a distance r from our moving charge is thus

$$\left(\frac{ev.\sin\theta}{r^2}\right)^2 \cdot \frac{u}{8\pi} \quad \text{or} \quad \frac{ue^2 \sin^2\theta}{8\pi\, r^4}.v^2.$$

The total energy of the magnetic field surrounding the particle is the sum of the energy in all these little volumes, from the surface of the particle away to infinity. The summation of this quantity does not present any difficulties. It is equal to $\frac{1}{3}\frac{e^2v^2}{a}$, where a is the radius of the sphere.

If m is the ordinary mechanical mass of the particle, the kinetic energy due to its motion will be $\frac{1}{2}mv^2$. In addition to this there is, as we have shown, a quantity of energy due to its charge. The total energy of the moving particle and its charge is thus

$$\frac{1}{2}mv^2 + \frac{1}{3}\frac{e^2v^2}{a}.$$

If we write this in the form $\frac{1}{2}\left(m + \frac{2}{3}\frac{e^2}{a}\right)v^2$, and compare it with the ordinary Newtonian expression for the kinetic energy of a particle $\left(\frac{1}{2} mv^2\right)$ we see that the particle behaves as if its mass m had been increased by an amount equal to $\frac{2}{3}\frac{e^2}{a}$. This is the extra or "electrical" mass, due to the fact that the particle carries a charge. It is evident that, even if m, the "mechanical" mass of the particle, is zero, the "electrical" mass due to the moving charge will still persist. Thus, even if our electron is a pure charge unassociated with any mechanical mass, it will still have an electrical mass equal to

$$\frac{2 \times (\text{electrical charge})^2}{3 \times \text{radius}.}$$

Since this "electrical" mass is really that of the magnetic field surrounding the particle, it resides not in the particle itself but in the medium surrounding it, that is, in that mysterious fluid which we call the ether. As soon, however, as we attempt to alter the motion of the particle this energy flows into it from all sides, so that, as far as experiments upon the particle itself are concerned, the results obtained are precisely the same as if it resided permanently there.

To make this somewhat novel idea a little clearer we may consider a close and very serviceable analogy, where the mechanism of the extra mass is a little clearer than in the electrical case. If any body is moving through water, or any viscous fluid, it carries with it a certain amount of the liquid through which it is moving. In the case of a sphere, for example, the quantity carried along by the motion of the body amounts to half the volume of the sphere itself. A long cylinder moving at right angles to its own length will carry with it a quantity of fluid equal to its own volume. On the other hand, if it moves in the direction of its own length the fluid entangled is practically *nil*. Thus, in order to set the body in motion with a velocity v, we have to supply to it energy enough to give this velocity, not only to the sphere itself, but also to the mass of fluid which it carries with it. That is to say, if M is the mass of the sphere itself, and M′ the mass of the attached fluid, the work done in starting the body is $\frac{1}{2}(M+M').v^2$. In other words, the body will behave as if its mass were increased by the mass of the fluid entangled by it. Just as in the electrical case, this extra mass resides in the surrounding medium.

It makes itself apparent at the surface of the sphere when the motion of the latter is altered in any way.

We may very profitably extend this analogy further in that part of our proof which still remains to be considered, by thinking of our electric field in terms of tubes of electric force. This conception we owe to the genius of Faraday. Untrained in the methods of mathematicians, he found the mere algebraical expression for the laws of electrical attraction both unsatisfying and unstimulating. He felt that if two bodies attracted each other it could only be because of some vital bond between them. Thus he imagined that every positive charge was the beginning of a certain number of tubes of force stretching out through space until, somewhere or other, they ended up on an equal but opposite negative charge.

By assuming that these tubes were in a state of tension so that they tended to shorten themselves as much as possible, and, further, that two similar tubes repelled each other, he was able to give a clear explanation of the various electrical phenomena. Thus the attraction between two oppositely charged bodies was the result of the tendency of the tubes connecting them to contract; the repulsion between two like charges was due to the mutual repulsion between the tubes of force radiating from each of them. An electric current was the motion of one of the ends of the tubes of force along the conductor, while the magnetic effect in the neighbourhood of the current was due to

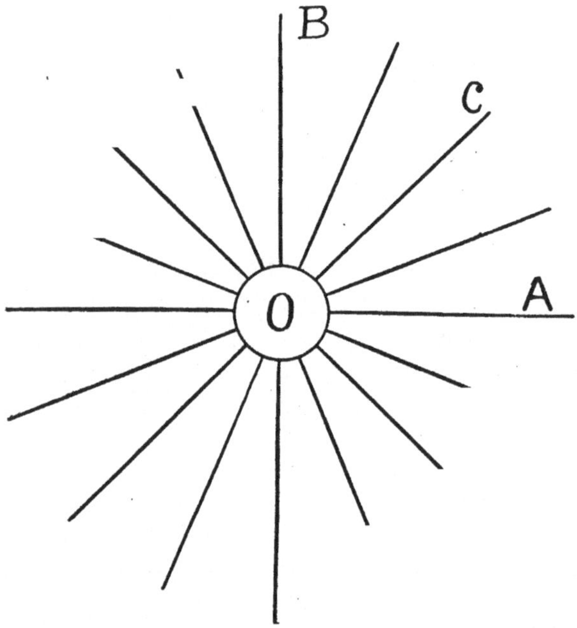

FIG. 16.—DISTRIBUTION OF TUBES OF ELECTRIC FORCE ROUND AN ELECTRON AT REST.

the motion of the tubes of force through the surrounding medium.

The concentration of attention upon the phenomena of currents which marked the latter half of last century tended to throw somewhat into oblivion this conception of Faraday. The recent researches with which we are dealing have, however, found in it a most powerful weapon, and are tending to give to these tubes of force a definite significance and concrete existence of which even their originator perhaps hardly dreamed.

Let us add to the postulates already made by Faraday this additional assumption, that a tube of force when moving through the ether is able to grip a certain quantity of that subtle fluid, in much the same way that a cylinder carries along with it a definite mass of any liquid through which it moves. For a single charged particle at a considerable distance from any other attracting bodies the lines of force will radiate out in straight lines from its surface, and, if the latter is spherical, the tubes will space themselves evenly around it owing to their mutual repulsion. In section, therefore, they will appear somewhat as in Fig. 16.

Suppose this system to be moving through space in the direction OA with a velocity v. If our analogy with the case of a body moving through water holds, the tubes, like OB, which move at right angles to their length, will grip the maximum amount of the ether through which they are moving, while tubes like OA, which are moving along their length, will carry none. It will thus be seen that the distribution of energy in the field surrounding the particle will correspond exactly with the distribution of energy in the electro-magnetic field of which it is supposed to be a visualisation. The electrical mass of the particle will then be the whole mass of the ether gripped by these radiating tubes.

Now it is a well-known fact that a cylinder moving through a fluid in the direction of its own length is not in stable equilibrium. In accordance with the well-known principle of least action, the cylinder will tend to turn so as to offer the greatest possible resistance to the motion, that is to say, it will tend to set itself at right angles to the direction in which it is being compelled to move. If this holds for tubes of force travelling through the ether, then tubes like OA are not in stable equilibrium, but will tend to turn into the plane of OB, which we may call the equatorial plane. It can be shown from the principles of electricity that such a tendency does indeed exist. The repulsion between the different tubes will, of course, tend to keep OA in position, and, as this repulsion is very great, at moderate

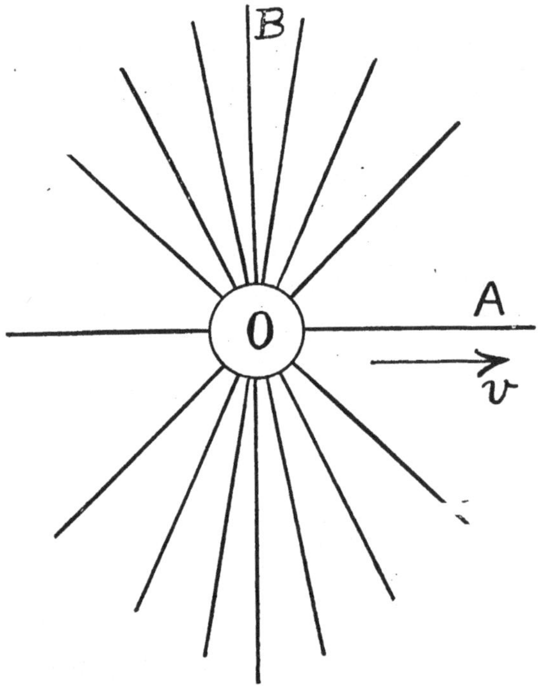

FIG. 17.—DISTRIBUTION OF TUBES OF ELECTRIC FORCE ROUND A MOVING ELECTRON.

velocities the effect we are considering will be quite negligible. The force, however, tending to displace the tubes increases with the velocity of the particle, so that at speeds approaching that of light there is a very perceptible shifting of the lines of force into the equatorial plane, as indicated in Fig. 17. But we have noted that the mass of ether gripped by the tubes in this position is greater than that which they originally held. Thus we arrive at the somewhat startling result that the electrical mass of a particle is not a constant, but depends, at any rate at high velocities, upon the speed with which it is moving.

We have followed out the theory of the variation of electrical mass from the standpoint of Faraday's tubes of force, because in the first place it gives us at once a clear and vivid picture of the processes which are going on in the field, while in the second place we are being gradually driven to the conclusion that some such entities do play a very important part in the phenomena of the electric field. The fact of the variation of electric mass with the speed, however, does not depend in the least upon the particular way in which we choose to regard the electric field, but is a direct though obscure consequence of the fundamental facts of the science.

The exact relation between the electrical mass of a moving charge and its velocity depends somewhat on the assumptions we make as to the distribution and geometrical configuration of the charge. Thus, starting from different ideas of the shape of the electron and the way in which its charge is arranged, different mathematicians have arrived at somewhat different formulæ for this relation. Fortunately, when these are reduced to figures the differences between the best of them are so small that our experiments are scarcely accurate enough to settle definitely between them. One of the most successful is that of Dr. Abraham, who deduces that

$$\frac{\text{electrical mass } m \text{ at a velocity } v}{\text{electrical mass } m_0 \text{ for small velocities}} = \frac{3}{4\beta^2}\left(\frac{1+\beta^2}{2\beta}\log\frac{1+\beta}{1-\beta} - 1\right),$$

where β is the ratio of the velocity of the particle to the velocity of light.

Let us see now how these results assist us to discover the nature of an electron. The total mass of the electron may be made up of two parts: its mechanical mass, which is constant and independent of the velocity with which the electron is moving; and an electrical mass, which increases with the velocity in the way we have just considered. If the electrical mass of the electron is small compared with its ordinary mechanical mass, then there will be very little variation in the total mass of the electron as the speed is increased. If, on the other hand, the whole of the mass is electrical in its nature, the total mass of the electron should increase at the rate given by the formula of Abraham. If part of the mass is mechanical and part electrical, the rate of increase of the whole mass of the electron with increasing velocity will be somewhere between these two values.

If we substitute numerical values in the equation given above, it will be found that the difference between m and m_0 is quite inappreciable unless the speed of the particle is very high indeed. Thus when the velocity is as much as one-tenth that of light, the increase in the electrical mass only amounts to about 1%, while even when the velocity reaches one-third that of light, that is to say, over 60,000 miles per second, the increase in the mass is only about 5%. As this is the maximum velocity attainable in a discharge tube it might have been thought that the experimental investigation of the variation of mass with speed lay forever beyond our reach.

Fortunately this is not the case. Experiments have shown that the β-particles from radium, while otherwise similar to cathode rays, travel with velocities enormously greater even than those of the particles in a discharge tube, in some cases approaching within 4% of the velocity of light itself. For speeds such as this the electrical mass should be nearly three times its ordinary value, and thus, by

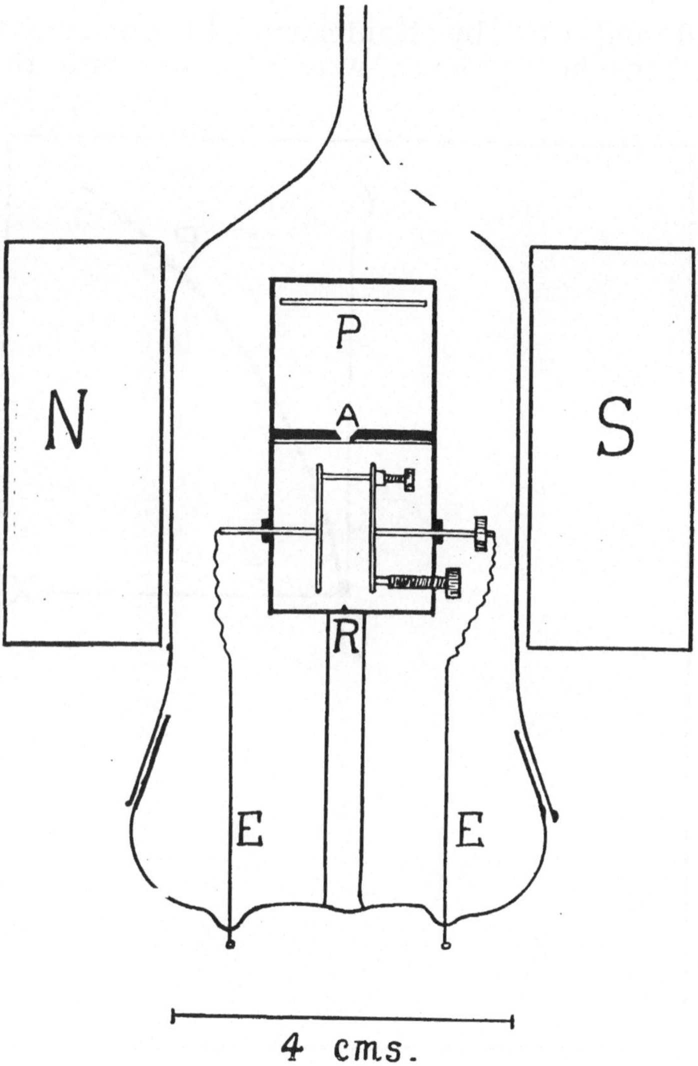

FIG 18.—KAUFMANN'S APPARATUS FOR DETERMINING THE MASS OF A β-PARTICLE.

careful experiments on these particles, we might well hope to put to the test the theories we have propounded.

Such experiments have been made with exquisite

skill and care by Kaufmann. In principle the methods he employed were identical with those

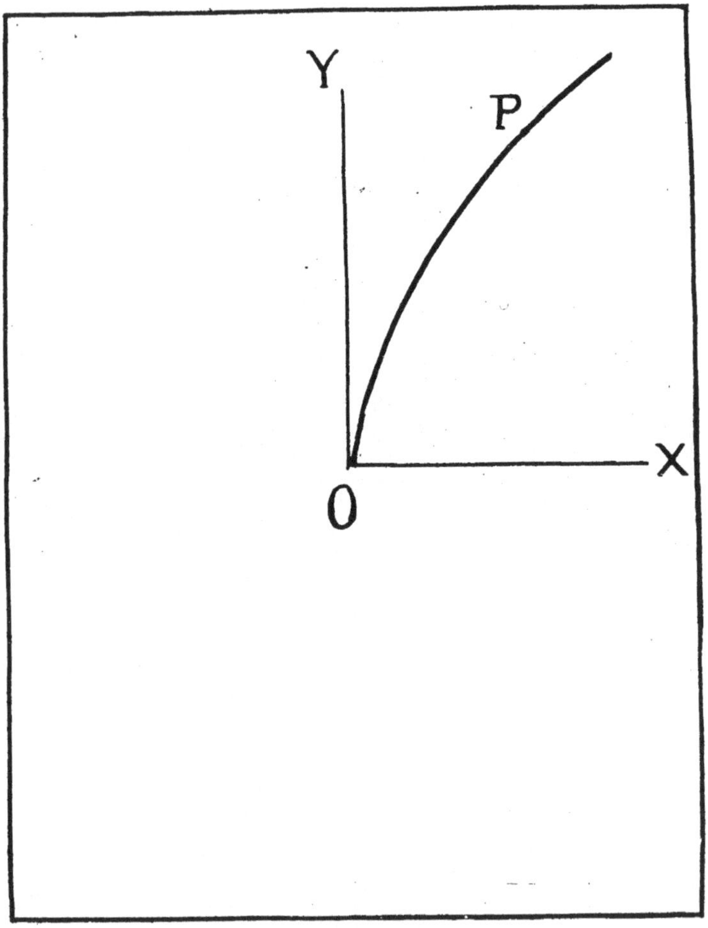

FIG. 19.—KAUFMANN'S EXPERIMENTAL CURVE FOR THE ELECTRIC MAGNETIC DEFLEXION OF THE β RAYS FROM RADIUM.

already described. The β-particles given out by a small quantity of radium placed at R (Fig. 18), passed through a fine hole in a lead screen at A, before finally falling on a photographic plate at P.

Before reaching the hole at A the particles passed between two parallel plates, which could be very carefully adjusted by means of a series of levelling screws, and which were charged to a high difference of potential. The whole of the apparatus was placed between the poles of a strong electromagnet N, S, arranged so that the electric and magnetic fields were parallel to each other and therefore the corresponding deflections they produced at right angles to each other, exactly as in Professor Sir J. J. Thomson's experiments on the positive rays. Apart from the differences in velocity all the β-particles are identical. Thus, for every value of the velocity v there is only one corresponding value for the mass m. Hence the beam of rays is drawn out on the plate into a single curve (Fig. 19), every point on which corresponds to particles with a particular mass and speed.

This curve can be measured up with a micrometer in exactly the same way as has been described for the positive ray curves, and the mass and speed of the particles deduced by equations (*c*) and (*d*) (Chapter IV.). The values of the constants k_1 and k_2 for the apparatus were very accurately determined, as were also the strengths of the electric and magnetic fields. In this way Kaufmann obtained results which he regarded as being accurate to within 1·5%, a very notable achievement.

Some of the results obtained are given in Table V. The first column of the table gives the velocity of the particles, the second the value of β in Abraham's formula, that is to say, the ratio of the velocity of the particle to the velocity of light. In the third column we have Kaufmann's experimental values for the ratio of the whole mass of the particles to their mass at small velocities. The last column gives the theoretical value of this ratio, deduced from Abraham's equation, on the assumption that the whole of the mass is electrical.

It will be seen how very closely, all things considered, the two sets of figures agree. The experimental values for the ratio m/m_0 increase, if anything, even more rapidly than theoretical. This would imply that, if anything, the mechanical mass of the electron is negative, which is, of course, an impossibility. The small discrepancy between the two sets of figures is no doubt due to a lack of

TABLE V.

Kaufmann's values for the variation of the mass of a β-particle with change of velocity.

Velocity of the β-particle.	$\beta = \frac{\text{velocity of particle}}{\text{velocity of light}}$.	Experimental value of m/m_0 for the β-particle.	m/m_0 for Abraham's equation.
2·88 × 10^{10} cm. per sec.	·963	2·70	2·52
2·85 " "	·949	2·35	2·24
2·80 " "	·933	2·09	2·14
2·65 " "	·883	1·96	1·81
2·49 " "	·830	1·70	1·61
2·41 " "	·801	1·61	1·54
2·20 " "	·732	1·41	1·41

knowledge as to the exact nature of the distribution of the electronic charge.

We have now discussed the evidence for our original statement that the whole mass of the electron is electrical in origin and due entirely to the charge it carries. Granting this conclusion, we can at once determine the dimensions of the electron, and so complete our knowledge of its properties.

We have seen that the electrical mass of a charge e for moderate speeds is equal to $\frac{2}{3}\frac{e^2}{a}$ where a is the radius of the space occupied by the charge. The mass of an electron is $8{\cdot}8 \times 10^{-28}$ gms., while the value of the charge it carries is $1{\cdot}57 \times 10^{-20}$ units. Substituting these values in the formula we find that the radius of an electron is $1{\cdot}87 \times 10^{-13}$ cms.

We have stated elsewhere that the radius of an atom is of the order of 10^{-8} cms., and have attempted to form some idea of the exceeding smallness of this magnitude. We may now say that small as the atom is, the electron is so much smaller that the electron bears to the atom which contains it very much the same relation as a pea to a cathedral.

We have seen that the whole of the mass of the electron is due to the charge which it carries. The thought at once suggests itself: Are there indeed two kinds of mass or is all mass electrical in its origin? Probably most physicists cherish this belief at the bottom of their hearts, but it cannot at present be said to be much more than a pious hope. The mass of a negative electron is about $\frac{1}{1700}$ part of the mass of a hydrogen atom. Neglecting the positive charge of the atom, of which we know practically nothing, it would require 1,700 electrons to make up the mass of a single hydrogen atom. This of course is not *a priori* an impossible number considering the smallness of the electron; and speculations along these lines were for a time freely indulged in. In this case, however, experiment failed to confirm the bold conjecture. The number of electrons in the atom has been determined at any rate approximately, and affords no support for such a theory.

As this number is one of considerable theoretical importance, we may perhaps be allowed to describe, very briefly, one of the methods by which attempts have been made to determine it. Consider a single electron flying through a sheet of metal. With the huge velocity which it possesses a sheet of aluminium, say $\frac{1}{5}$th mm. thick, is not a very serious obstacle to a β-particle. Neglecting for a moment the positive charge which keeps the atom together, the various atoms which our particle encounters must present to the particle the appearance of a series of large chambers, empty save for a certain number of particles, minute in comparison with the atom, and similar to itself.

At a considerable distance these fixed electrons will have little effect on the moving corpuscle. If, however, in its flight it approaches one of them closely it will be repelled, as the two particles each carry a negative charge. The nearer it approaches the greater will be the repulsion, and the further the moving particle will be deflected from its original path. We may call this a collision. It is, in fact the same sort of collision as we imagine to take place between two gas molecules. The direction in which the particle will be deflected will be perfectly arbitrary, and a second collision may actually undo the work of the first. On the whole, however, the greater the number of collisions the greater will be the deflection, and it is not beyond the possibilities of mathematics to determine the most probable effect of a given number of these deflections. The problem is indeed mathematically equivalent to the well-known theorem of the drunkard's walk. Suppose a drunkard walks so many yards before falling down, and after each fall, having lost all sense of locality, sets out again in some new and arbitrary direction, at what distance from his starting point will he be most likely to be found after a given number of falls? This problem, under a somewhat different title, has been solved by Lord Rayleigh in that marvellous compendium of mathematical physics, the "Theory of Sound."

If then we can find experimentally the most probable deflection of a β-particle after passing through a certain thickness of solid matter, we can deduce the number of collisions it has made in its passage through the solid, and hence the number of electrons in unit volume of the substance. Since the number of atoms per cubic centimetre of any substance can readily be deduced from the data given in Chapter III., the number of electrons in each atom can at once be determined.

A series of experiments upon these lines was recently performed by the author. By various experimental devices, which it would take us too far from our course to describe, the mean angle through which an originally parallel pencil of β-rays was scattered in passing through sheets of various substances was determined. By experimenting on sheets of different thicknesses the author was able to show that the mathematical treatment outlined above was applicable to this particular case, and hence to deduce the number of electrons in the atoms of the various elements experimented upon.

The matter is, perhaps, hardly so simple as we have described it. We have neglected so far the effect of the positive electrification which we know must be present in the atom if the latter is to be electrically neutral. This will exert a certain attraction on the flying electrons.

Unfortunately, we are not yet acquainted with the nature of positive electricity. Professor Sir J. J. Thomson's experiments on the positive rays, brilliant as they have been, have not at present thrown much light upon this exceedingly difficult problem. For the present the term "positive electrification" remains for the physicist very much what the term "catalytic action" is for the chemist

—a not too humiliating method of confessing ignorance. If we suppose that the positive electricity is distributed uniformly over a sphere the size of the atom (a hypothesis which lends itself very readily to mathematical treatment), the author's result would indicate that the number of electrons in an atom is almost exactly three times its atomic weight. That is to say, the number of electrons in a hydrogen atom would be three. If we go to the other extreme, and suppose that the positive electrification is a sort of nucleus at the centre of the atom, and that the electrons revolve round it somewhat after the manner of the rings of Saturn, the number of electrons in a hydrogen atom works out at unity, the number in any other atom being equal to its atomic weight. The assigning of unit atomic weight to hydrogen would then have a very definite physical significance, as it would be the lightest atom which could possibly exist. In either case the number of electrons in an atom is only a very small multiple of its atomic weight. We cannot, therefore, assign any appreciable fraction of the mass of the atoms to the negative electrons it contains.

There still remains, of course, the possibility that the mass is electrical, but that it resides in the positive portion of the atom. If the formula for the electric mass be examined, it will be seen that for a given charge the mass is inversely proportional to the radius of the sphere upon which it is concentrated. If we suppose that the positive charge on the hydrogen atom to be concentrated upon a sphere only $\frac{1}{1700}$ of the size of the negative electron, its mass would be 1,700 times as great, that is to say, equal to that of the hydrogen atom. Our perfect ignorance of the nature of positive electricity renders the suggestion not untenable, though evidence for it is sadly lacking. For the present our belief in the electro-magnetic nature of all mass remains an expression of our faith that all the varied phenomena with which we have to deal are manifestations of some single principle or essence which underlies them all. To what extent future experiment will realise our belief in the present instance remains to be seen.

CHAPTER VI.

The Chemistry of the Model Atom.

We have now learned all that is to be known about electrons, their nature, their mass, their charge, and their dimensions. We have seen that they form a part of all atoms, and have discovered, at any rate approximately, how many of them we may expect to find in an atom of any given element. We have seen also that electrons form a very minute part of the mass, and a still smaller part by volume of the whole atom.

It might be thought therefore that their influence in determining the properties of the atom would be proportionately small. There is much evidence however that this is not the case, as we shall have occasion to see during our further survey of molecular p[illegible]. The electrons are small but they are mobile. It is entirely owing to this mobility that we are so well acquainted with their properties. It is owing to this same mobility that they exercise such a marked influence upon the chemical and physical character of the atom.

Our difficulties in constructing a satisfactory model of an atom lie, as we have already seen, in our almost complete ignorance of the nature of the residue of an atom from which the electrons have been extracted. The atom can lose at any rate some of its electrons without any serious change of properties, and as the normal atom is neutral, the residues after extracting the electrons must possess positive electrification. Beyond this we know little or nothing.

Three views are possible. In the first place we may assume of our residues what the earlier atomists assumed of their atoms, that they are fundamentally different in character, that copper is copper, and tin is tin, and that there is no more to be said about it. This would be to abandon all hope of attaining any explanation of those recurring similarities between the properties of the different elements which are so clearly brought out in that greatest of the generalisations of descriptive chemistry, the periodic classification of the elements. This is a policy of despair.

Assuming, then, that the remainder of all the atoms is composed of different amounts of the same stuff, call it matter or positive electrification as we will, this substance may be either continuous or divided up into larger or smaller particles. The fact that the α-particles given out by all radioactive substances are identical might suggest that these particles are the stones from which the edifice of the atom is built up. The α-particle is, however, too complex a structure to serve as a protyle ; it is, for example, four times as heavy as our lightest atom, and, in all probability contains at least two electrons. Professor Nicholson has built up an interesting chemical system from protyle atoms of atomic weight, 1, 2, 3, and 4, but these in themselves would in all probability be complicated structures, and in the absence of further evidence on which to work it seems better to proceed at once to the fundamentals.

The existence of a positive electron is not quite beyond the realms of possibility. It may be that there is some sort of undetected one-sidedness about our experiments which accounts for our complete failure to observe it. If at any time it makes its appearance, it is possible that something of the argument which follows may have to be referred to the positive rather than to the negative electron.

Taking all the facts into consideration, the most probable hypothesis at the present moment is the one which regards the positive part of the atom as forming either a uniform sphere, coterminous with the atom, or a dense nucleus at its centre. In the first hypothesis the electrons would move about within the positive sphere and would be attracted with a force varying directly as their distance from the centre of the sphere. On the second hypothesis some, at any rate, of the electrons would be outside the positive charge and revolve as satellites around it, like the rings round the planet Saturn, being attracted to it in this case with a force varying inversely as the square of the distance.

The latter suggestion has very much to recommend it, but it has also the very great defect from our immediate point of view that it does not lend itself at all readily to mathematical treatment. In other words we cannot easily determine from the laws governing such a system the way in which the electrons would be arranged, and hence we cannot easily deduce the properties to be expected in such an atom. It is most probable that the arrangement of the electrons in the two cases would exhibit the same general features, though, of course, the actual values in any given case would be different. As it is not at all likely that either of these simple systems will eventually be found adequate to represent the atom of any real substance, we may, without any loss of generality, confine our attention to the simpler case.

The grouping of a number of electrons in a sphere of positive electrification which is just sufficient to neutralise their charges at points outside the sphere, has been investigated theoretically by Professor Sir J. J. Thomson. A single electron if at rest will place itself at the centre of the sphere, or if in motion will describe a circle around that point with a radius depending on its velocity. Two electrons would occupy the opposite ends of the diameter of a circle whose radius is equal to half the radius of the positive sphere while three electrons would arrange themselves at the points of an equilateral triangle. The arrangement of four electrons at the corners of a square is, however, not a stable grouping if the ring of electrons is at rest, but becomes possible if they are revolving round the centre with a speed which exceeds a certain critical value. If their velocity falls below this the system collapses into a tetrahedron with an evolution of energy.

The determination of the arrangement of any number of electrons in a space of three dimensions becomes at first difficult and finally impossible. A few cases have, however, been worked out. Thus six electrons would arrange themselves at the corners of a regular octahedron, but, on the other hand, the arrangement of eight at the corners of a cube is not stable and will collapse into two tetrahedra, one inside the other, unless we station at least one other electron within it. We are thus brought face to face with one of the most important generalisations in electron grouping, namely, that it is quite impossible for any system to make a large display of electrons near its surface without a corresponding stock within. The presence of a given number of electrons near the outside of the atom demands that there shall be a certain definite number of electrons within in order that the system may be stable. To use a commercial metaphor, an atom cannot have all its goods in the shop window. Otherwise there will be inevitable collapse into some less pretentious system.

The analysis becomes simpler, and leads to the same general results, if we consider the electrons as confined to a single plane instead of being able to move in three dimensional space, a supposition which is not altogether without justification on experimental grounds. This case has the further

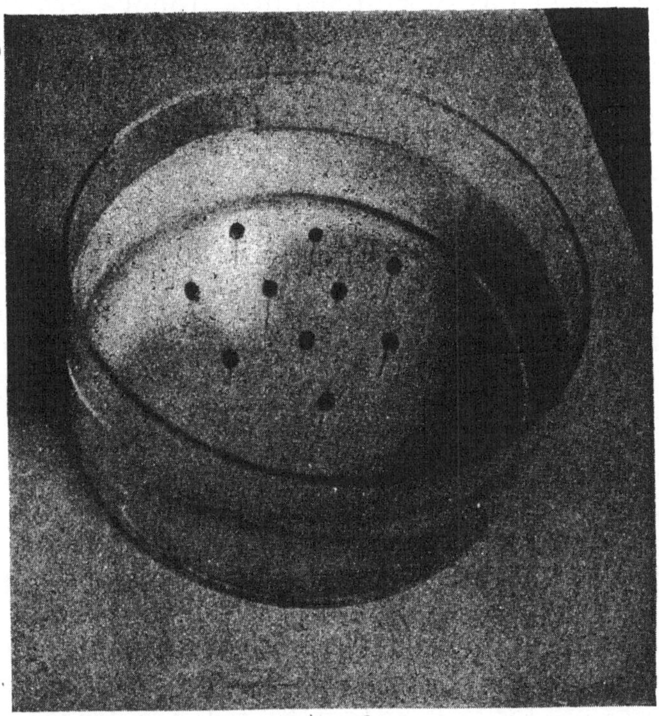

FIG. 20.—A "MODEL" ATOM.

advantage that it can be studied experimentally in a variety of ways. Of these the very beautiful experiments of Mayer, though designed for quite other purposes, are perhaps the simplest to reproduce.

A large number of needles are magnetised together in a solenoid, and are floated vertically in a basin of water by pushing all their (say) north poles into small corks, their south poles being at the same depth below the water. These latter repel each other, just as do the electrons, according to an inverse square law. To represent the action of the positive sphere we may place a strong electro-magnet beneath the bowl with its north pole upwards. It can be shown that the attraction of this magnet for the tiny south poles is (for the horizontal plane in which alone they are able to move) approximately proportional to their distance from a point immediately above the pole. Fig. 20 is a photograph showing ten of these small magnets. It will be seen that they are arranged in two rings of seven and three respectively. The attracting electro-magnet is concealed below the board.

On making the experiment it is found that the greatest number of magnets which we can have in an empty ring is five. If a sixth is added the ring breaks up, the magnets gradually settling down into an arrangement of five on the outside with one in the middle. The number which must be placed in the middle rapidly increases with the number in the outer ring. Thus an outer ring of twelve requires at least eight inside, and one of thirty no less than one hundred and one for stability to be possible. As these inner magnets are subject to exactly the same laws as the outer ones, the whole system, whether of magnets or electrons, splits up into a series of concentric rings, with a perfectly definite number of particles in each.

A complete mathematical solution has been obtained by Professor Thomson. Table VI. contains his results foı the model atoms in which the number of electrons is not more than fifty-two. I would point out very emphatically at this point that no stress must be laid on the exact numbers given in this Table. The simplifications we have made are too many to render it even possible that these arrangements represent actual atoms of any real substance. Neither does the weight of the argument rest on any numerical relations between the numbers of electrons in the different rings. The general law of electron grouping enunciated above when discussing three dimensional systems is amply sufficient for all that follows, and is all we have at present any right to make. It is, however, of the greatest assistance in following the arguments to clothe our abstract principles in concrete form and to consider some definite case where the relationships can be expressed in numerical form. The system chosen is the simplest for the purpose.

The first row in Table VI. contains, with these limitations, the "atoms" for which there is only one ring of electrons, the second those with two, and so on, the upper number being in each case that contained in the outer ring. Thus 11, 5, 1, which commences series C, implies that this atom contains an outermost ring of eleven electrons with an inner ring of five and a single electron at the centre. Remembering that the number of electrons in the atom is proportional to the atomic weight, it will be seen that this Table represents a series of model elements arranged according to their ascending atomic weight.

A very cursory glance at this Table at once reveals a very striking similarity to Mendeleef's periodic classification of the elements. Thus, for example, if we assign any chemical or physical property to the existence in the atom of an arrangement of electrons such as (5, 1) this property will disappear as we pass along the series to atoms with different groupings, and consequently different properties. Proceeding still further, however, we come to a place where the original grouping (5, 1) is restored with the addition of an extra outer ring of eleven electrons (11, 5, 1).

This grouping again disappears as we pass along series C, to reappear again with an extra ring of fifteen at the beginning of series D, and with a further ring of seventeen in series E.

Thus any property associated with a given grouping of electrons will recur again and again at intervals, as the number of electrons in the model atom is increased. It can hardly be necessary to point out how closely this resembles the arrangement of the elements according to atomic weight where the properties of the element sodium, for example, disappear as we proceed to elements of higher atomic weight, only to make their appearance again with slight modifications when we arrive at the element potassium.

There are many other resemblances which may be pointed out between the periodic Table and our scheme of electrons. In the first place it will be seen that in the same group the atoms of series D and E have four rings in common with each other, while those in series B and C have only two. The elements of higher atomic weight will resemble each other more closely than the elements of lower atomic weight. This again is a commonplace of the periodic classification. Further, it will be seen that at certain places in the Table the groupings are in rapid change, while at others the configuration alters much more slowly. This again corresponds very closely with what actually occurs in the arrangement of the elements where starting, for example, at sodium, the alkaline properties gradually disappear as we pass through magnesium and aluminium, to be replaced by acid ones at sulphur and chlorine. As we proceed further, however, there is a sudden recurrence of the alkaline properties at the next element, potassium.

Again, although the groupings shown in the Table are all stable, they are not the only possible arrange-

TABLE VI., SHOWING STABLE ARRANGEMENTS OF ELECTRONS IN A POSITIVE SPHERE OF ELECTRICITY.

								2	3					4		5		
A								2	3					4		5		
B	5 1	6 1	7 1			8 1	8 2	8 3		9 3		10 3	10 4		10 5		11 5
C	11 5 1	11 6 1	11 7 1		12 7 1	12 8 1	12 8 2	12 8 3	13 8 3	13 9 3		13 10 3	13 10 4	14 10 4	14 10 5	15 10 5	15 11 5
D	15 11 5 1	15 11 6 1	15 11 7 1	16 11 7 1	16 12 7 1	16 12 8 1	16 12 8 2	16 12 8 3	16 13 8 3	16 13 9 3	17 13 9 3	17 13 10 3	17 13 10 4	17 14 10 4	17 14 10 5	17 15 10 5	17 15 11 5
											l	m	n	o	p	q	r
E	17 15 11 5 1	18 15 11 6 1	18 15 11 7 1			etc.											
	s	t															

ments of electrons. Thus, while we cannot have a ring of six electrons unless there is at least one electron inside it, we may conceivably have more than this without disturbing the equilibrium. Thus an atom with ten electrons might be represented either by (8, 2) as in the Table, or by the grouping (7, 3). The latter grouping is the one actually seen in the photograph (Fig. 20). In general, one of these arrangements would be more stable than the other, that is to say, its energy would be greater and it would be more likely to survive. At certain points, however, the different configurations might be so evenly balanced as to be almost equally stable. In this case we should have a series of elements of almost exactly the same atomic weight, but differing somewhat, though probably not markedly, in properties; a case corresponding very closely to the puzzling series of rare earths which occur in the middle of the Periodic Table.

It may be asked, what properties can be associated with a grouping of electrons in a sphere of positive electrification? If we are on the right track, we may expect sooner or later to be able to interpret every phenomenon associated with the atom in terms of such a system. But this is a far cry, and it must be confessed that our present knowledge does not extend very far in this direction. We can, however, suggest several important properties which may, and one or two which can be proved to be due to causes of this kind.

In the first place, we may almost certainly include all those forces of cohesion, adhesion, and the like which go by the name of molecular: those attractions which, quite insensible at any appreciable distance, are yet at distances comparable with the diameter of a molecule strong enough to rivet together the atoms and molecules in a solid so firmly that in some cases it requires stresses of many tons to the square inch to drag them asunder. Since these molecular forces determine tenacity, elasticity, and melting point for solids, and surface tension and vapour pressure for liquids, it will be seen that a very considerable and important set of properties are thus brought within the scope of our hypothesis.

The most probable and, indeed, at present, the only explanation of these molecular attractions, is that they represent the resultant force between the different parts of the two electrical systems which make up the atoms, each of which is neutral as a whole. Let us take as an illustration a very simple concrete case. Suppose the black dots in Fig. 21

FIG. 21.—DIAGRAM TO ILLUSTRATE COHESION.

represent the outer rings of electrons in two adjacent atoms, and that the effect of the positive electricity around them may, for the sake of simplicity, be regarded as concentrated in the white circles between them.

Since each of the atoms, as a whole, is electrically neutral, the attractions and repulsions between the different parts at any distance which at all greatly exceeds the diameter of the atom will be vanishingly small. If, however, as in the illustration, the two atoms are brought so close together that their distance is only a fraction of this diameter, this is no longer the case. It can readily be shown that the atoms will turn so that a negative electron in the one is faced by the positive portion of the other. There would thus be an attraction between them which would increase with great rapidity as the distance between the atoms was diminished. The law of force would thus be similar to that actually found for these molecular forces.

The magnitude of the forces obviously depends on the number and arrangement of the electrons in the atom. A complete theory of cohesion is not possible in the present state of our knowledge. There is little doubt that it will be found to proceed on lines not very different from the explanation suggested above.

Although our methods of analysis do not enable us to pursue this problem further, another very important characteristic of the atom, namely, its emission spectrum, has proved more amenable to investigation. The evidence obtainable from a study of spectral lines is so important that we shall deal with it more fully in a later chapter. For the present it may be noted that, before the electron had been isolated and its properties determined by direct experiment, it had been shown by Lorentz that the systems emitting the vibrations which constitute ordinary light were negatively charged particles, and further, that the ratio of the charge to the mass was of the order of $1 \cdot 7 \times 10^7$. It was also shown that these same electrons were the cause of the absorption of light and its dispersion in transparent media. Thus, practically the whole range of optical phenomena of the atoms are due to the electrons which they contain.

This point is so important that we shall return to it in a separate chapter. In the remainder of the present chapter we will confine our attention to that important, fascinating and mysterious quality of the atom which determines its chemical properties, namely, valency.

Ever since the time of Berzelius, the view has

been held that the forces binding the different atoms together within the molecule were electrical in their origin, that, for example, the molecule NaCl was formed by the mutual attraction of a positively charged sodium atom for a negatively charged atom of chlorine. This view was, of course, based on the phenomena of electrolysis. When an electric current is passed through a solution of sodium chloride in water, it is found that the sodium makes its way to the negative electrode, the chlorine to the positive. Thus, the salt behaves as if the sodium atoms bore a positive charge and the chlorine a negative. The conclusion seems irresistible, that the force binding together these opposites in the same molecule is the electrical attraction between their charges.

It was found necessary, from consideration of the electrolysis of such compounds as $CuCl_2$, etc., to assume that the atoms of certain elements carried charges amounting to twice, three times, etc., the charge carried by a sodium or chlorine atom, and thus the notion of valency arose, the valency of the element being the number of times its charge in the ionic state exceeded that of sodium or chlorine. For elements such as carbon and nitrogen, which are never found in the free ionic state, the valency was deduced from their chemical analogy to elements of known valency. We may say, then, that positive valency represents the tendency of the neutral atom to part with its electrons, while negative valency is the tendency of the neutral atom to absorb electrons into itself from without. If electro-positive and electro-negative atoms are present together, these tendencies can both be satisfied by the mutual interchange of one or more electrons. The passage of an electron from one atom to the other will make the former positively the latter negatively charged, and the attraction of the positive atom for its lost electron will constitute on our hypothesis the force of chemical combination between the two atoms.

Although these ideas in one form or another are as old as Davy and Berzelius, chemists have made, up to the present, little direct use of them, finding the notion of bonds of affinity sufficient for all practical purposes. So long as the electric charge was regarded as something apart and distinct from the atom itself, something of a different nature superadded to it by external means, the electrical theory, except in so far as it explained the phenomena of electrolysis in which it had its rise, was no more fundamental than the theory of a mysterious chemical bond.

With the electron theory of matter, the situation is quite changed. It is evident that the question whether a given atom will tend to lose or gain electrons, and how many electrons can be so lost or gained, will depend upon the number and arrangement of the electrons within it, and the forces with which they are retained in the atom. If we can determine these, and if we can show that they lead to a similar distribution of valency with atomic weight to that actually demanded by chemical considerations, we shall feel that we have made a very hopeful start in the solution of this fundamentally important question.

In the first place, it must be pointed out that all the electrons in the atom are not held in the same way. The majority of the electrons are too firmly held to be displaced by anything short of a disruption of the whole atom. A certain number, however, are held comparatively loosely. Optical experiments have shown that the number per cubic centimetre of the comparatively mobile eelctrons which determine the dispersion of the substance is not many times greater than the number of atoms present. Similar results are obtained from a consideration of the numbers of free electrons in a metallic conductor. It will obviously be these free electrons that determine the valency of the atom. It seems very probable that the greatest number of these that any atom can gain or lose does not exceed eight, and it may be noted that this is the greatest charge ever found to be carried by a positive particle.

Since valency is a property of the simplest atoms, it seems probable that the valency electrons form the innermost ring of the atomic system. Let us suppose, now, that the maximum number of electrons which can be contained in this inner shell is eight, the minimum, of course, being zero, and let us suppose, further, that there is a tendency in every atom to give this inner ring either its maximum or minimum value. An atom with an inner ring of five electrons would then either part with its five valency electrons, in which case it would have a positive charge and, consequently, a positive valency of five, or attract three more electrons from without, making up its complement to eight. In this case it would be electro-negative in character with a charge and a valency of three. It will be noted that the sum of the two valencies is equal to the maximum number of particles in the inner ring, namely, eight.

Which of these two actions would take place would depend on the nature of the atoms with which it was placed in contact. If surrounded by atoms which tended to part with electrons, it would be electro-negative, if by atoms with a strong attraction for free electrons, it would eventually give up some, or all of its five to the stronger atom and remain electro-positive with a maximum valency of five. The atom would correspond closely to such a substance as phosphorus, which forms compounds such as PH_3, in which it shows a negative valency of three, while, with the strongly electro-negative element, chlorine, it forms compounds such as PCl_3 and PCl_5, in which it has a maximum positive valency of five.

To compare our results with the periodic table, we must suppose the elements of Group I., lithium, sodium, etc., to possess, when in their neutral state, only one valency or free electron, which they give up with the greatest readiness. The elements of the next group should have two free electrons and a positive valency of two. At the same time, there would be the possibility of their taking in sufficient electrons to bring their total up to eight, in which case they would possess a possible maximum negative valency of six. I am not aware that any compounds have been prepared in which this valency is exercised. It might, in any case, be expected to be very feeble.

The number of valency electrons in the atom, then, increases uniformly as we pass along the series, until we reach the halogens with a total number of seven. These have a possible positive valency of seven, as they might conceivably lose the whole number they possessed. This valency is probably exercised in compounds, such as I_2O_7. As they contain very nearly the maximum number possible, they are, however, much more likely to acquire the one necessary electron to make their total up to eight, and thus act as strongly electro-negative elements, as in the great majority of their stable compounds, *e.g.*, NaCl, we know they do. From energy considerations we should expect that the number of electrons in the inner ring would tend to that stable number to which it was nearest to begin with. Thus, elements possessing fewer free electrons than four would be electro-positive, those with more would tend to make their complement up to eight, and be electro-negative. The elements of the carbon group (Group IV.), with four valency electrons would be almost indifferently positive or negative; as we find them to be, both CO_2 and CH_4, being stable compounds. Our complete valency scheme is thus represented by Table VII., where the Roman numerals indicate the group in the periodic classification. It is most interesting to note that Professor Abegg, in a very illuminating paper on valency, arrived at a precisely similar scheme from purely chemical considerations. He suggested that every element had two valencies, a normal and a contra valency, the normal being supposed to be the stronger. The positive valencies were the normal

TABLE VII.—TABLE OF VALENCIES OF THE PERIODIC GROUPS.

Group.	I.	II.	III.	IV.	V.	VI.	VII.
Normal valency .	+ 1	+ 2	+ 3	± 4	− 3	− 2	− 1
Additional or contra-valency .	− 7	− 6	− 5		+ 5	+ 6	+ 7

for elements of Groups I. to III., the negative for Groups V. to VII., Group IV. being indifferently plus or minus. The sum of the normal and contra valencies for any element was eight. It will be seen that this suggestion is exactly embodied in our theoretical table.

The phenomena of variable valency and valencies of different signs have no difficulties on our hypothesis. It is, in fact, exactly what we should expect. It is not necessary that the whole of the available electrons should be extracted in one step. It is evident that the work done in extracting an electron from the atom will be greater for the second than for the first, and it is quite easy to imagine circumstances under which the ionising force was sufficient to produce the expulsion of one, but not of two electrons, while under more favourable conditions, both might escape.

The inert gases of the helium group possess no valency and, hence, presumably, no free electrons. We have already remarked that the phenomenon of the dispersion of light is due to these free electrons, which are thus sometimes spoken of as dispersional electrons. It is interesting to note, in support of the hypothesis we have advanced, that the dis-

TABLE VIII.—NUMBER OF FREE ELECTRONS IN THE ATOM DEDUCED FROM THE DISPERSION OF LIGHT.

Substance.	Calculated Number of Free Electrons.	Number Assumed (maximum positive value).
Hydrogen . . .	1·4	1
Calcium	1·5	2
Carbon	3·7	4
Silicon	3·9	4
Oxygen	4·4	6
Chlorine	6·2	7

persion of light in helium is abnormally small, while that of the halogen gases, which, on our theory, possess the maximum number of valency electrons, is very large.

Drude, by means of a rather abstruse and somewhat approximate calculation, has deduced from these considerations the number of free electrons in the atoms of different elements. Some of the results are given in Table VIII. Remembering that the number of electrons in the atom must be an integer, and that the calculations are only approximate, his calculated values seem fully to bear out the assumption we have made above.

Again, our hypothesis explains directly the curious connection between the sign of the valency and electrical conductivity. With the exception of hydrogen, every electro-positive element conducts electricity when in the solid state. This fact is such a commonplace, that an explanation, perhaps, at first sight, hardly seems necessary. On any other theory of valency, however, it would be difficult to account for this invariable co-existence. On the assumption, however, that electro-positive elements are those which easily give up electrons, it is evident that since these electrons will be free to move under the electric field, these elements will be able to conduct electricity, while, since the electro-negative elements tend to absorb electrons, there will be few electrons present in them in a free state, and since the atoms themselves are not able to move through the solid, the current will find no carriers, and so will be unable to pass.

One further direct deduction from our hypothesis is of interest. It can be shown that when two atoms of different sizes come into contact, the attraction of the smaller atom for the electrons contained in the larger exceeds that of the larger atom for the electrons in the smaller. There is thus a resultant force tending to drive the electrons from the larger to the smaller atom. It has been shown that in the simplest possible case an electron will actually pass between two similar atoms if the radius of the smaller is less than seven-tenths that of the larger.

Thus, among atoms of the same type, those with a smaller volume will tend to acquire, those with a larger volume will tend to lose, electrons. In the same group, therefore, the elements of greater atomic volume will be more electro-positive (or less electro-negative, as the case may be) than the elements with the smaller atomic volume. In any one group of elements the atomic volume increases with the atomic weight. Hence, we arrive at the general result, that the electro-positiveness of the elements in any group increases with increasing atomic weight. This relation has often been pointed out. A very striking example of it is seen in Group V. of the periodic classification, where the lightest element, nitrogen, is usually electro-negative, while the heaviest, bismuth, is generally electro-positive.

We have now traced out some of the consequences of the hypothesis that valency is determined by the number of loosely bound or valency electrons contained in the atom, these electrons being identical with those which determine the dispersion of light. We have also made the additional assumption that the maximum number of such electrons which can be contained in any atom is eight, and that in every case there is a tendency to make this number either eight (negative valency), or zero (positive valency).

This latter hypothesis has, it must be confessed, somewhat of an *a posteriori* air about it, and it may be interesting to investigate, briefly, whether there is any property of electron grouping from which this apparently arbitrary assumption can be deduced.

Qualitatively, at least, we can show that some such quality is a direct consequence of the laws of electron groupings which we have formulated above. It was shown that for any ring or shell of electrons to persist, there must be contained within it a certain definite number, at least, of other electrons.

Let us suppose now, that for an outer ring of A electrons this number is a, while for a ring of A + 1 electrons it is b. Consider an atom with a ring of A electrons having n other electrons within it where n is some number greater than a, but less than *b*. By suitable means we may extract from this atom a number of electrons equal to $n—a$, without destroying the equilibrium of the outer ring A. If, however, we succeeded in extracting one further electron, the system would become unstable and the atom would collapse.

Now we know that any real atom is a very stable structure indeed. Certain atoms, it is true, owing to circumstances we shall have to consider later, spontaneously break up into simpler substances, but there is no known agency by which we can bring about the dissolution of an atom of ordinary matter. The phenomena of radio-activity have furnished us with the clue to our failure : the energy required is too great. On the other hand, the energy required to extract single electrons from a neutral atom has been determined, experimentally, and is triflingly small. A single α-particle, for example, has sufficient energy to produce many thousands of ions. Thus, the work required to reduce the number of electrons in the atom below the critical value *a* would be out of all proportion greater than that necessary to extract the electrons above this number. In effect, the value *a* would be a barrier which we could not overstep.

Similarly, we could add to the atom sufficient electrons to make its number up to *b*, without bringing about any change in atomic structure. Beyond that point, a further electron would necessitate an outer ring of A + 1 electrons, with a rearrangement of the atomic structure.

Thus, the two critical values *a* and *b* constitute, as it were, barriers which we cannot pass. We may extract $n - a$ electrons, at most, giving the element a positive valency of $n - a$, or we may add $b - n$ electrons, giving the atom a negative charge and negative valency of this amount. More than this we cannot achieve. The sum of these two valencies is obviously $b - a$, which is a constant for all elements with a ring of A electrons.

We have presented the argument in general terms, rather than from a consideration of the model scheme in Table V., because we did not wish to suggest that it depended in any way upon the numerous simplifications and approximations which were necessary in calculating that table. This granted, however, a concrete example taken from that table may serve to make the argument more clear.

Consider the elements with an outer ring of seventeen. We have lettered them from *l* to *s* in Table VI. The number of electrons within the ring in element *m* is twenty-six, which is one more than is absolutely necessary for the stability of the ring. Thus, the element *m* might lose one, and only one electron without any violent rearrangement of the system taking place. Moreover, as the system is not much more than stable, the forces retaining the electron within the atom are not very strong, and this electron would be very easily displaced. The element *m* would thus be monovalent and strongly electro-positive. The next element, *n*, could lose a total of two electrons without derangement of the system. It would thus be divalent and electro-positive, but as it would naturally be more stable than *m*, its activity would be less marked. Thus, as we pass along the series, the positive valency increases numerically while, at the same time, it decreases in intensity.

At the other end of the series the element *r* could take in one extra electron without raising its total number sufficiently to necessitate an outer ring of eighteen. It would thus be monovalent and electro-negative. Theoretically, it might also lose six electrons and be hexavalent and electro-positive. From energy considerations, however, this would only occur in exceptional circumstances. Similarly, the element *q* preceding it would be divalent and electro-negative, though less actively so than *r*. Thus, our series of model atoms gives us a valency system very similar to that already described.

In conclusion, it may be noted that the electron theory of valency, while more fundamental and complete, has yet a very close resemblance to the "bond" theory, and is thus capable of including in itself all the very important and interesting results which have been achieved on that hypothesis. For consider an electro-positive atom which has lost one

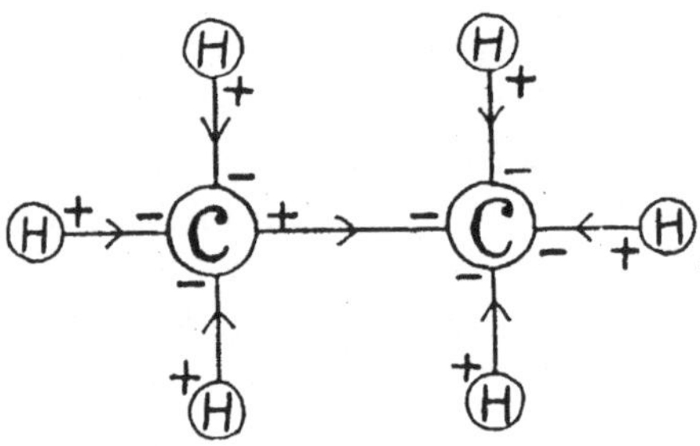

FIG. 22.—STRUCTURE OF ETHANE.

electron. It will be the starting point of a Faraday tube of force, which, when a compound is finally formed, will end up upon the extra electron contained in the electro-negative partner of the combination. Similarly, a divalent atom would be the starting point of two such tubes, and would require two monovalent or one divalent atom for its saturation. Thus, we may say that the chemical bonds represent the Faraday tubes linking up the different atoms in the molecule.

There is this difference, however, between the two theories, that while a bond is not usually regarded as implying direction, a Faraday tube is essentially different at its two ends, and implies a finite difference of potential. Take, now, the case of ethane, C_2H_6. Regarding all the hydrogen atoms as positive, three Faraday tubes have their negative ends upon each of the carbon atoms (Fig. 22). If the bond between the carbon atoms is a Faraday tube it must run from one atom to the other. The atom on which it starts will then have a total negative charge of 2.*e*, while the other has a total charge of 4.*e*. The two carbon atoms are thus not in precisely the same electrical state.

No change in properties has yet been observed, attributable to this difference. In the case of

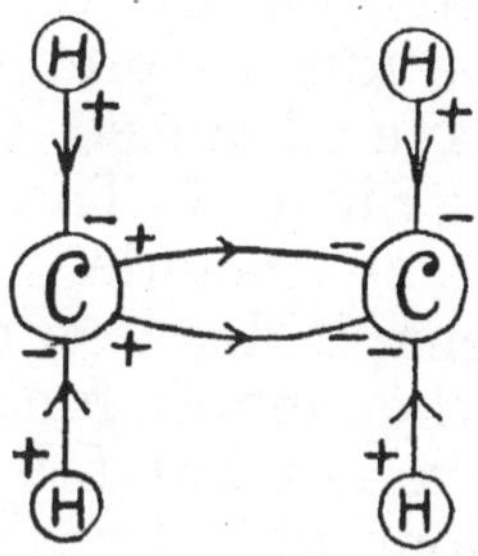

FIG. 23.—STRUCTURE OF ETHYLENE.

ethylene, however (Fig. 23), where the difference between the atoms is more marked, one carrying a charge of 4.*e*, while the other is electrically neutral, such differences have been noted. Take, for example, the property of the refraction of light. It has been found that the refraction produced per molecule by many organic compounds can be calculated by assigning certain definite values to the atoms of carbon, hydrogen, and the rest, the molecular refraction of any compound being then equal to the sum of the values for all the separate atoms which go to make up its molecule. It has been found, however, that if the carbon atom is present in the ethylene linking, a certain additional amount must be added in for every atom so linked up. Thus, the atomic refraction of the carbon atom is apparently not the same in its neutral as in its charged state. A precisely similar result is obtained if we consider other additive properties, such as molecular volume. Thus, there seems to be a distinct difference in the properties of the carbon atom in these two electrical conditions.

We have developed this subject of valency at considerable length, because it is one of the most fundamental, and, at the same time, has been one of the most mysterious of all atomic properties. While it would be idle to pretend that the hypothesis which we have advanced, which is due, in the main, to Professor Sir J. J. Thomson, is at present entirely adequate to include the whole of these puzzling phenomena, it does, on the whole, correlate so successfully the more fundamental facts of valency that we cannot but feel that, at any rate, we are working along right lines and may look with some confidence to a complete solution in the perhaps not very distant future.

CHAPTER VII.

THE ATOM IN VIBRATION.

IT can easily be shown that when a moving electric charge is accelerated or retarded in any way, a wave of electro-magnetic disturbance radiates out through the surrounding space. Consider an electron with its Faraday tubes moving along in the direction AB (Fig. 24). If the electron is moving with uniform

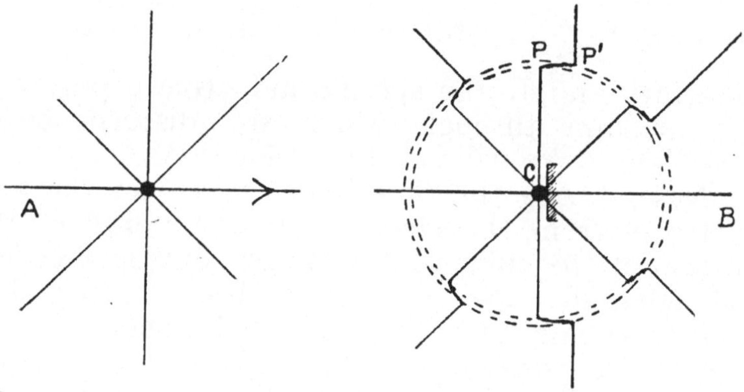

FIG. 24.—PRODUCTION OF ELECTRO-MAGNETIC DISTURBANCES BY THE STOPPAGE OF AN ELECTRON.

velocity these tubes will travel along with the electron as if they were rigidly attached to it. Suppose, however, that this electron on reaching a point C is suddenly stopped in its career. The ends of the Faraday tubes attached to the electron are suddenly brought to rest.

Now a Faraday tube is in some respects very like a stretched rope. For instance, it has a tension along its length, and possesses, as we have seen in a previous chapter, mass. Thus, just as when one end of a rope is jerked, a certain definite time elapses before the displacement reaches the other end, so a displacement in a Faraday tube does not reach all points of it simultaneously, but is propagated along it with a definite velocity. In the case of the stretched rope the velocity is equal to $\sqrt{\frac{T}{m}}$ where T is the tension on the rope and m its mass per unit length. On similar principles it can be shown that the velocity of a disturbance along a Faraday tube is equal to $\frac{1}{\sqrt{\mu K}}$ where μ is the magnetic permeability and K the specific inductive capacity of the medium through which the disturbance is travelling. We will call this velocity V.

Now, during the time it has taken for the jerk to travel from the electron O to a point P, the portion of the tube beyond P, not having received the signal that the electron has been stopped, has been travelling on with its original velocity and is now in the position P′, where PP′ is the distance which the electron would have gone in the time taken for the disturbance to travel along the tube from O to P. Thus a kink is developed in the tube which travels outward along it with the velocity V.

The bent portion of the tube PP′ is moving perpendicular to its length and will thus, by the principles we have already developed, produce a magnetic field in the medium which will also travel out with the kink. If we consider all the tubes surrounding the particle, it will be seen that these bent portions will form a kind of sheet of disturbance travelling outwards as an expanding spherical shell or wave.

If we can determine for any medium the values of μ and K in the same system of units, we can calculate the velocity V with which the disturbance is propagated in that medium. It will be remembered that the electrostatic system of units was obtained by defining K the specific inductive capacity of air as unity. The electro-magnetic system on the other hand assigns this value to μ the magnetic permeability of air. The comparison of the two systems can best be performed by measuring the capacity of the same condenser according to both systems of units. This operation is described in most text-books of electricity, and its details are outside the scope of our present work. The results obtained are, however, most striking and important. The result of a long series of experiments at the American Bureau of Standards extending over several years has given a value for V of $2{\cdot}997 \times 10^{10}$ cm. per second. This is, within the limits of experimental error, identical with the best determinations of the velocity of light, which have given a mean value of $2{\cdot}998 \times 10^{10}$ cm. per second. The agreement between the values obtained for the two velocities is too close to be accidental. It affords a convincing proof of the electro-magnetic nature of light.

In the case we have considered where the particle is suddenly brought to rest a single pulse is formed the thickness of which is equal to the distance through which the disturbance can travel in the time taken to stop the particle. Such pulses are given out when cathode rays in a discharge tube impinge on a solid anticathode. They give rise to the well-known phenomena of Röntgen or X-rays.

If instead of stopping the particle dead we cause it to oscillate about a mean position, a series of undulations will be set up in the Faraday tubes attached to it, such, for example, as are seen when one end of a long rope is shaken regularly to and fro. Thus a series of electro-magnetic waves move out from the vibrating electron, their frequency being equal to the number of vibrations made by the particle per second. Their wave length is equal to the distance travelled by the disturbance during the time of one complete vibration. If this lies between ·0004 and ·0008 mm. these waves will produce on our eyes the impression of light. This, in its crudest form, is the Faraday-Maxwell electro-magnetic theory of light.

Not only may light originate in the vibration of electrons; all our experiments drive us to the conclusion that such vibration is the only way in which light can be produced. Uncharged matter is apparently quite incapable of setting up any kind of disturbance in the æther. It is the electric charge which serves as the only link between the two.

We can, however, proceed further than this. Modern research has enabled us to identify with certainty the vibrating systems which emit the myriad lines in the spectra of the elements with the electrons, with the properties of which we are now growing familiar.

Michael Faraday, with almost uncanny prescience, seems to have felt that the relation between magnetism and light must be of the closest, and sought for it with unremitting diligence. It is one of the little ironies of life that the only effect which he did discover (the magnetic rotation of the plane of polarisation) was one of the very few phenomena which the theory even now is hardly adequate to explain. The effect he sought for was not discovered until, more than thirty years later, his experiments were repeated with stronger magnetic fields and far more powerful methods of spectrum analysis by Zeeman.

He found that if a source of light was placed between the poles of a strong electro-magnet and the light emitted was examined spectroscopically, a curious change took place in the appearance of the different lines in the spectrum, each being split up or decomposed into two components when the light was viewed along the lines of force in the magnetic field, or three when viewed at right angles to this direction.

Let us consider briefly the effect of applying a magnetic field to the charged particles emitting the vibrations as they oscillate about their mean positions. In general the path described by the particle will be somewhat complicated. We can, however, resolve its motion in the usual way into

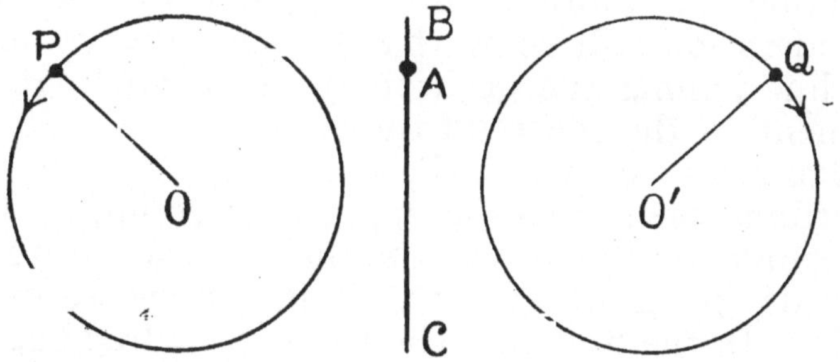

FIG. 25.—ILLUSTRATING THE DECOMPOSITION OF A HARMONIC MOTION INTO TWO CIRCULAR VIBRATIONS.

vibrations parallel to and at right angles to the lines of force in the magnetic field. The vibrations performed along the lines of the field will be unaltered by the application of the field since the latter produces no mechanical effect on a particle moving in its own direction.

It will be quite different with those vibrations which are executed in directions at right angles to this. It is shown in works on mechanics that any harmonic vibration can be regarded as the sum of two circular motions of properly chosen amplitude and phase. For example, the harmonic motion of a particle A vibrating along the straight line BC

(Fig. 25) is exactly equivalent to the motions of two equal particles P and Q describing the circles shown in the diagram with equal velocities but in opposite directions.

It will be sufficient, therefore, and will lead to no loss of generality if we consider the charged particles as describing circles at right angles to the magnetic field, under the action of forces directed towards the centres of the circles. Since the motion of the charged particles is now everywhere at right angles to the magnetic field there will be a mechanical force acting on the particles which will be everywhere at right angles to the magnetic field, and to the direction of motion of the particles. A little consideration will show that it must always act in the line joining the position of the particle at any moment to the centre of the circle.

Particles such as P and Q are, however, describing their circles in opposite directions. Thus the magnetic force will be in one case directed towards, in the other away, from the centre of the circle. In the first case the total force dragging the particle towards the centre is increased ; the circle is contracted and the time of vibration becomes more rapid. In the latter the effective force is diminished, with the result that the orbit expands and the vibrations become slower.

An apple whirling round at the end of a string will furnish a very useful analogy. If the tension of the string is increased the apple is drawn in and its revolutions become more rapid. If, on the other hand, the tension is relaxed, the apple takes a wider sweep, while the time it takes to complete a single revolution becomes longer.

Thus, instead of all the particles describing their orbits in exactly the same time and thus producing light of one frequency only, one set have been accelerated, the other retarded by the magnetic field, so that instead of a single line the spectroscope will reveal two, one on each side of the original position of the line. By optical means we can distinguish between clockwise and counter clockwise rotations, and can thus determine which of the two has been accelerated and which retarded by the magnetic field. Since the effect on a positive particle would be exactly the reverse of that on a negative, we can thus determine the sign of the

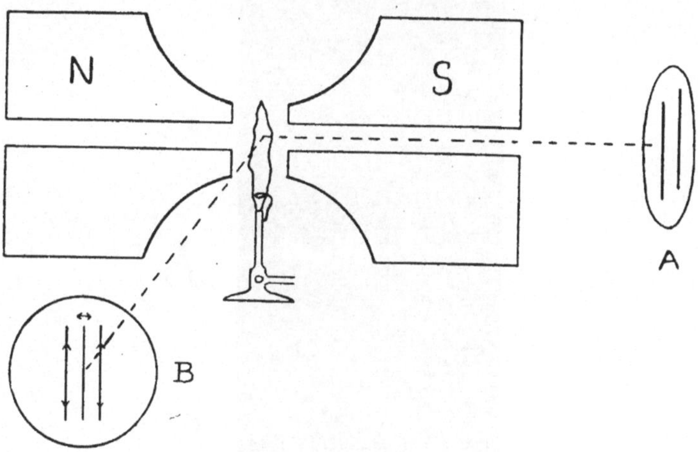

FIG. 26.—ILLUSTRATING THE ZEEMAN EFFECT. N, S ARE POLES OF ELECTRO-MAGNET. A, B REPRESENT THE APPEARANCE OF A SINGLE LINE EMITTED BY THE FLAME AS SEEN THROUGH A SPECTROSCOPE PLACED IN THE TWO POSITIONS.

charge carried by the vibrating particle. In every case so far examined it has been found to be negative.

If we view the flame along the lines of force, these two new lines are the only ones seen, as the vibrations taking place in the line of sight produce no effect on the eye. If, however, we look across the field, the vibrations executed along the lines of the field will also produce their effect in the spectroscope, and since their period is unaltered by the field they will give a line in the position of the original undecomposed line. Since all the circular paths are being now viewed end on, all three lines will appear to be plane polarised. These results are shown diagrammatically in Fig. 26.

This in its simplest form is exactly the effect discovered by Zeeman in 1896. He found that if a source emitting a line spectrum were placed between the poles of a very powerful electro-magnet

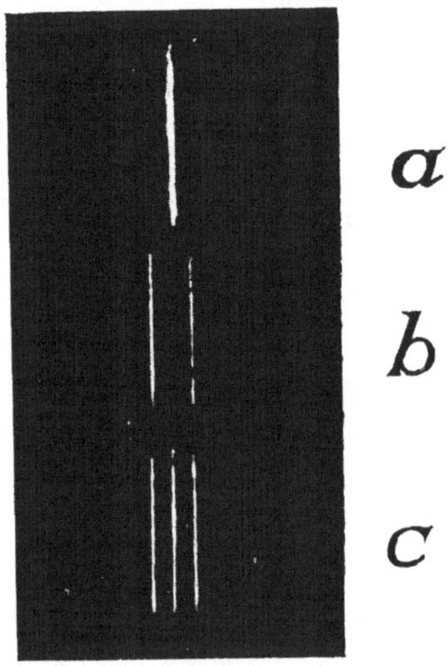

FIG. 27.—THE ZEEMAN EFFECT. (*a*) UNDECOMPOSED LINE. (*b*) SAME LINE VIEWED ALONG THE LINES OF FORCE OF A MAGNETIC FIELD. (*c*) LINE VIEWED AT RIGHT ANGLES TO LINES OF FORCE.

and viewed with an interference spectrometer of high resolving power the line was split up into three components when viewed at right angles to the magnetic field and into two when seen along the lines through a hole in the pole piece. This is very well shown in the photograph reproduced in Fig. 27. When, at the suggestion of Lorentz, who at once saw the correct explanation, the polarisation of the different components was examined, it was found to agree exactly with that described above.

We have seen that these observations have shown that the vibrating system is negatively charged. We can, however, go further than this. Following up the argument given above it can easily be shown* that if λ_a and λ_c are the wave lengths of the two outer components into which a single line of wave length λ_0 is split up

$$e/m = \frac{\lambda_c - \lambda_a}{H \cdot \lambda_0^2} \cdot 2\pi V$$

where e/m is the ratio of the mass to the charge for the particles and V is the velocity of light. Now Runge and Paschen found that for a certain series of mercury lines, for example, the value of $\frac{\lambda_c - \lambda_a}{\lambda_0^2}$ was 2·14 for a magnetic field of 24,600 units, while V is of course 3×10^{10} cms. per second. Substituting these values in the equation we have

$$e/m = \frac{2 \cdot 14 \times 2\pi \times 3 \times 10^{10}}{24{,}600} \text{ or } 1 \cdot 65 \times 10^7.$$

Considering that the displacement produced by even the strongest fields amounts generally to somewhere about one-fifth of the distance separating the two D lines of sodium, it will be seen that this result agrees even better than might have been expected with the value of the same ratio $1 \cdot 77 \times 10^7$ deduced for the negative electron from experiments on cathode rays. The identity of the light-emitting systems with the cathode ray particles must be regarded as established beyond all challenge.

Many spectral lines show the Zeeman effect in this simple form. Others, however, give more complicated effects, some of the lines of Molybdenum, for example, splitting up into seventeen and even nineteen components. We have, in assuming that

* See Appendix C.

the vibrating electron executed harmonic vibrations, neglected the possibility of its motion being disturbed by the forces due to neighbouring electrons in the same atom. If these are not negligible complications ensue, and the time of vibration is no longer independent of the direction in which it takes place. The matter has been discussed at length for certain simple systems of electrons by Professor Sir J. J. Thomson, and his paper may be found in the *Proceedings* of the Cambridge Philosophical Society, Vol. 39, by any who are interested in pursuing the matter more deeply.

The lines in the spectra of an element thus represent the different modes of vibration of the electrons connected with it. It might seem that with a complete knowledge of these periods at our disposal it should be possible to reconstruct the atom with certainty. Theoretically, no doubt, this should be so. The mathematical difficulties are, however, appalling, and only in the case of a few very simple systems is the analysis at all possible. Thus the results to be looked for from spectroscopy are at present qualitative rather than quantitative.

Even so, however, a study of spectra is most luminously suggestive. A careful analysis of the innumerable lines which are dotted apparently at random through the spectra of the elements has shown that they are by no means the confused jumble of periods which at first sight they appear to be. It has been found that in very many cases they can be grouped into definite series, every line of which can be obtained by substituting particular whole numbers for a variable quantity "*m*" in a general equation the form of which is the same for all elements.

This was first pointed out by Balmer for the hydrogen spectrum, but it was afterwards shown by Rydberg that Balmer's equation was merely a particular case of a very general law which could be stated so as to apply to all the elements giving series spectra. He showed that if n is the frequency of the vibrations corresponding to a given line (the frequency is the number of vibrations made per second by the particle emitting the light) then the frequency of all the lines in a given series can be represented by the equation—

$$n = n_0 - \frac{N}{(m + \mu)^2}$$

the different lines composing the series being obtained by putting m equal successively to the whole numbers, 1, 2, 3, . . . N is a constant so universal in its character that it applies not merely for all the series of a given element but for all the series of all elements giving series spectra at all. The constants n_0 and μ differ for different elements and are characteristic of the particular series.

That this method of classifying a spectrum is no mere mathematical fiction is very clearly brought out when the character of the lines making up a single series is closely observed. It is found that all the lines of a single series are exactly similar in appearance. Thus all of them appear sharp and distinct with definite clear-cut boundaries, or else they are all indistinct and diffuse. The two types of lines are never associated in the same series. Again, any physical change which we bring about in one of the lines of a series is reproduced simultaneously in all the others. Thus, all the lines split up in the same way under the action of a magnetic field. If one of the lines is broadened by the application of pressure to the gas emitting the spectrum, all the others are broadened to a similar degree, and when the method of producing the spectrum is altered, all the lines have their brightness either enhanced or diminished at the same time. We can hardly avoid the conclusion that they are all different modes of vibration of the same system.

An analogy may make this clearer. It is well known that when an organ pipe is sounded it gives out in addition to its fundamental note a series of overtones, the fullness and richness of the organ notes being due to the presence of these extra tones. In such a simple system the overtones are always harmonics, that is to say, their frequencies are always exact multiples of the lowest mode of vibration. In mathematical language we can say that the different frequencies are obtainable by substituting various whole numbers for the quantity m in the equation, $n = m \cdot n_o$.

In more complicated vibrators the relationship becomes more complicated while at the same time the importance of the overtones becomes even more pronounced. In the case of a sounding bell the relation between the different modes of vibration is so complicated that it has not yet been completely worked out.

This is not all, however. Not only do the spectra of a single element show these close numerical relationships. They are also found to exist between the spectra of different elements belonging to the same group of the periodic table. The spectra of the different elements of the alkali metals, for example, are so closely connected that it is actually possible to calculate the whole spectrum of, say, rubidium or cæsium, from that of potassium, and a knowledge of the relative atomic weights of the three elements. The spectrum of cæsium is practically the same as that of potassium but is shifted bodily towards the violet end of the spectrum. We are thus driven to the conclusion that the vibrating systems in the two cases are identical in form, and differ from each other only in being placed in slightly different environments.

In terms of our model atom, if we regard the series spectra as being given out by the inner rings of electrons, these rings are identical for the three elements, but in the case of the two latter the extra outer rings added on to increase the atomic weight produce some alteration in the external forces acting on the rings and so cause a shift in the absolute frequencies of the vibrations without altering the relation between them. In the words of astronomical theory, the outer electrons perturb the vibrations of the inner ring without destroying their character. Or, returning to our analogy of the bell, it is the same bell which is sounding but in a somewhat different medium.

One difficulty, a very formidable one, remains, and may, perhaps, already have occurred to the reader. The spectra of many of the elements is exceedingly complicated, containing in some cases many hundreds of well-defined lines. A contemplation of the iron spectrum wrung from the great American spectroscopist Rowland the exclamation that, compared with an atom of iron, a grand piano must be a very simple affair indeed.

We have seen, however, in a former chapter that the number of electrons in an atom is, on the whole, not very large. In all probability the hydrogen atom contains only one electron, and certainly not more than three. A single electron has only three degrees of freedom and therefore only three modes of vibration, and in the simple atom these three would all have the same period and thus coalesce into a single line. As a matter of fact, the ordinary spectrum of hydrogen contains a fair number of lines while what is known as its secondary spectrum is very complicated. The difficulty seems at first sight almost insuperable.

To solve the riddle, Professor Sir J. J. Thomson has hazarded the bold but almost certainly correct conjecture that the lines in the spectrum of hydrogen and, indeed, of many other elements are not emitted by the atoms of the element at all but by systems which only exist when the gas is thrown into a luminous condition. When we consider the absolute identity of the spectrum of an element whether situated on the earth or away on the furthest star of which we have spectroscopic data, this hypothesis may seem to be the most startling we have yet advanced. It is, however, not without a very considerable experimental basis.

Consider for a moment the phenomena known as selective absorption of light, of which the well-known experiment of the reversal of the sodium lines is a good example. If white light is passed through sodium vapour enclosed in an iron tube and the transmitted light examined with a spectroscope there can be seen across the ordinary coloured spectrum two sharp, dark lines, corresponding exactly in position with the two bright lines, which would be seen if the sodium were raised to incandescence. This is the cause of the dark Fraunhofer lines which are to be seen crossing the spectrum of the light from the sun.

These effects are due to a form of resonance. The sodium atom contains systems of electrons which, when suitably excited, give out light of the frequency corresponding to the two sodium lines. To use an analogy from sound or from wireless telegraphy the sodium atom is tuned to this frequency. It will therefore begin to vibrate violently if oscillations of this period fall upon it. In doing so it abstracts the energy of the radiation in tune with itself, so that the latter is very rapidly absorbed, while light of frequency remote from this passes on almost unaffected. Thus, after passing through a very small thickness of material the light is robbed of all the constituents which have frequencies corresponding to those of the radiating systems within it. Conversely all the electronic systems which have natural periods falling within the limits of the visible spectrum must produce a dark band when white light is passed through them.

Now, it is remarkable that such absorption lines are not seen when white light is passed through a

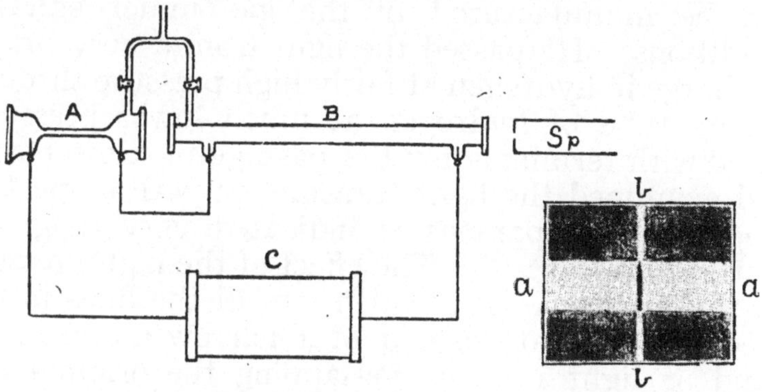

FIG. 28.—LADENBURG'S EXPERIMENT.

A, discharge tube containing hydrogen at high pressure. B, discharge tube containing hydrogen at low pressure. C, induction coil working both tubes. *Sp*, spectrometer for viewing the light transmitted. The inset shows the appearance seen in spectrometer when the tubes are working. *aa*, broadened band due to hydrogen in A. *bb*, spectrum due to the tube B alone. The dark line where the two bands cross shows the reversal of the line by the systems formed in B by the discharge.

column of hydrogen, or indeed of most other simple gases. There are, for example, no dark lines to be seen in the spectrum when examined after passing through a long tube of hydrogen gas, corresponding to, say, the brilliant lines of the hydrogen spectrum in the red or in the blue. Indeed, the laws of the absorption and dispersion of light in hydrogen are so simple that they can be explained on the assumption of a single absorption band far away beyond the violet end of the spectrum. This is, of course, what we should expect from the simple nature of the hydrogen atom. There is thus no escape from the conclusion that the vibrating systems which emit the hydrogen spectrum are not present in the gas in its normal condition.

A very interesting experiment made not long ago by Ladenburg has shown, however, that they can be manufactured in the gas under suitable conditions. He passed the light from a very bright discharge in hydrogen at fairly high pressure through a long tube of hydrogen at much lower pressure, fitted with terminals for the passage of a discharge, and examined the light transmitted with a spectroscope. His apparatus is indicated very diagrammatically in Fig. 28. The effect of the high pressure on the discharge is to widen out all the lines in the spectrum so that, instead of a narrow line a broad band of light is seen overlapping the original line on each side. When there was no discharge passing through the long tube of hydrogen there was no trace of any absorption of any part of this broadened band.

If, however, a feeble discharge was sent through this long tube of gas it was seen that the centre portion of the broad line corresponding to the original line and hence having the period of the light emitted by the tube at low pressure became distinctly less bright than the outer portions, showing that the long column of feebly glowing gas was now absorbing from the light emitted by the brighter hydrogen tube those wave lengths corresponding in period to the light it was itself giving out. There are some difficulties in the experiment, and a certain synchronism of the two discharges seems to be necessary for its success. On the whole, the conclusion seems irresistible. The hydrogen line spectrum is emitted by systems of electrons which are not present in the gas in its normal condition, but which are formed when the gas is thrown into a luminous state.

Let us consider for a moment the ways in which an atom can be made to emit luminous vibrations. Experiments have shown that for an atom to emit its characteristic spectrum it is necessary that it should be in a region of intense ionisation or recombination. It seems that nothing short of driving an electron into or out of the atom will suffice to set in vibration the light emitting systems of electrons. It is obvious that this condition is fulfilled when the spectrum is excited electrically, either by sparking between terminals of the substance or by a low pressure discharge through a gas. It is interesting in this connection to note in the latter case that experiments have shown that, in the dark portions of the discharge the ionisation is very feeble, while it is most intense in the brightest portions of the positive column. The ordinary method of producing a spectrum by placing the substance in a Bunsen flame is also merely a rough-and-ready way of producing ionisation in the substance, Professor H. A. Wilson having shown that a Bunsen flame, especially if it contains salt vapours, is the seat of very intense ionisation.

Consider an atom placed within this ionised space. It meets the ionising agent and an electron is ejected, leaving the atom with a positive charge. The loss of this electron will alter the forces within the atom and the remaining electrons will oscillate about their new positions of equilibrium before finally settling down. These vibrations, if of suitable periods, will set up luminous radiations and give rise to a set of lines in the spectrum. The positively charged atom is now surrounded by a cloud of negative electrons which fill the surrounding space. It attracts them and, normally, one would enter to take the place of the lost member. If however, the mutual kinetic energy of the two is sufficient the electron will revolve round the attracting atom without falling into it in exactly the same way as the earth revolves around the sun. The revolutions of this electron, being isochronous, will give rise to a single line in the spectrum with a period corresponding to the time taken by the electron to describe the circle.

If the law of force between the atom and the electron were the simple one of the inverse squares, an orbit of any size would be possible under suitable conditions. Taking into account the fact that the electron is losing speed owing to the fact that it is all the time radiating away its energy in the form of light waves, it will be gradually drawn nearer and nearer the atom so that its path would be a sort of spiral ending up in collapse into the atom. Since slightly different times would be occupied in describing each part of the curve the spectrum thus produced would be continuous.

The neighbourhood of the electrons within the atom and their repulsion of the external one would complicate the simple law of force, and in general the attraction of the atom for its satellite would depend not only on the distance but also on the direction of the line joining the satellite to the atom. The actual law of force would thus depend upon the number and arrangement of the electrons within the atom, just as much as if the electron were revolving inside it.

Sir George Darwin, in his monumental work on Periodic Orbits has shown that, if the single system of a satellite revolving round its planet is disturbed by the presence of a third body, the number of orbits which the satellite may describe is no longer infinite, but that there are certain regions through which the particle must not pass if it is to continue to revolve as a satellite around the planet.

In the simple case considered by Darwin this space did not quite enclose the planet, but in more complicated cases such as we have to deal with when we come to a consideration of the atom, he was of the opinion that it might very well take the form of a series of closed rings surrounding the atom. In this case there would be a series of separate rings in which alone the electron could revolve, and each of which it would describe in a definite time peculiar to that ring. Each of these rings would thus give rise to a separate line in the spectrum. Since the atom would only attract one electron at a time each atom would at one time be only emitting one of these extra lines.

These systems would, of course, not be permanent, as the electron would constantly be losing energy by radiation. After a time its energy would fall below that necessary to keep it in rotation in the ring it was describing, and it would collapse through ring after ring, until finally it was reabsorbed into the atom. Fresh systems would, however, be constantly being formed in the ionised gas, and the supply would be kept up as long as the exciting conditions were maintained.

The return of the lost electron into the atomic fold would obviously stir up fresh vibrations among the electrons already present. Since the atom is now neutral and not positively charged as it was when the electron left it these vibrations would not have the same periods as those set up when the electron was expelled. There is some evidence for the suggestion that the vibrations corresponding to the return of the electron form the band spectrum of the element, while those set up on its expulsion, together with those due to the special systems, make up the different series of line spectra.

Very recently an ingenious Danish mathematician, Dr. Böhr, has worked out mathematically a theory of spectra very much on the lines described above, and has in this way obtained a formula for the spectral series identical in form with the empirical formula of Rydberg which we have already discussed. He has from his theory deduced an expression for the universal constant N of Rydberg's equation in terms of the mass and charge on an electron and other known physical quantities, which gives a numerical value for that quantity within about 5 % of the value obtained from direct spectroscopic observations. His assumptions will require further consideration before they can be unreservedly adopted. They appear to involve the abandonment not merely of Newtonian mechanics, but also of the ordinary Faraday-Maxwell electro-magnetic theory. The closeness of the agreement between the results of the theory and of experiment is, however, at least remarkable, and seems to show that, at any rate in its general outlines, the theory of the origin of spectra which we have propounded above is a step in the right direction.

We have, however, wandered far from the solid certainties of the Zeeman effect and its consequences, with which we commenced this chapter, into the dark and mysterious hinterland where science is in the making. In a few more years we may hope that this will be transformed into plain and open ground. At present we must pause here, lest we lose ourselves altogether in a quagmire of fascinating but groundless speculations.

CHAPTER VIII.

THE MOLECULAR THEORY OF MATTER.

WE have now surveyed very briefly what is known and surmised of the nature and properties of the atom and molecule. In the brief space of the present chapter let us indicate one or two of the many points of interest which arise when we come to consider the molecule, not as standing by itself, but as massed together in myriads to form a fragment of what we know as matter.

In solids, we must regard the molecules as fixed relatively to each other, and we have in an earlier chapter considered briefly the nature of the molecular forces, cohesion, adhesion, and the like, which bring about this result. But though the molecules are fixed, in the sense that they are not free to move about in the solid, we cannot consider them as being altogether motionless. Every body, unless it is at the absolute zero of temperature (a temperature which has not yet been reached in any experiment, although the brilliant researches of Kamerlingh Onnes have brought us within 4° of it), contains a certain definite amount of heat. Now heat is merely energy in a molecular form. The work which we do in raising the temperature of a substance is in reality spent in increasing the energy of the individual molecules of the substance, and thus in every body which is not at the absolute zero, the molecules must be in some sort of motion, the rapidity of which increases as the temperature is raised.

The molecules of the solid are not free to move from their positions, but two forms of motion are possible to them. They may either spin round like a top, or oscillate backwards and forwards about their mean position, like the bob of a pendulum, or like a heavy weight suspended by an elastic cord.

In the case where the molecule is a system made up of two or more atoms, both these kinds of motion take place. The atoms spin round their common centre at a rate which increases as the body gets hotter, while at the same time the molecule as a whole, oscillates backwards and forwards about its mean position with ever-increasing amplitude. If, however, the molecule consists of a single atom, only the latter form of motion seems to be possible, all the evidence going to show that thermal energy is unable to set a single atom spinning about its own centre.

For the sake of simplicity we will consider the case of the non-atomic molecule, where the thermal energy is all used in the second kind of motion. As the body gets hotter and hotter, the molecule moves faster and faster, and the swings which it takes on each side of its normal position get larger and larger. The molecular forces which keep it in position, however, fall off very rapidly as the distance between the neighbouring molecules is increased, while at the same time the force required to retain the molecule in position increases with every increase in velocity. Hence, sooner or later, a point is reached when the molecular forces are no longer able to retain the atoms in position. The solid loses its cohesion and its rigidity. In other words it melts.

It is obvious that melting sets in when the temperature is just sufficient to give the molecules the necessary minimum of velocity to overcome the molecular forces retaining them in position, that is to say, it will be a constant for any definite substance. Also as work must be done to break down these inter-molecular bonds it is evident that energy must be supplied to the body in the form of heat to enable it to melt. This is the latent heat of liquefaction.

Though the molecule has broken its old bonds, it is still within the range of molecular attraction of many of its neighbours, and has hardly escaped from one cluster before it is attracted into another. Its rate of progress is thus comparatively slow. In time, however, any given group of molecules will be found dispersed throughout the whole of the liquid and thus diffusion is brought about.

Consider now the state of affairs at the surface of the liquid. Every molecule as we know attracts every other molecule which may happen to be within a certain distance from it which we denote as the sphere of molecular attraction. In the body of the liquid, this attractive force is more or less neutralised by the fact that the molecule we are considering is surrounded on all sides by others, all pulling in different directions. Hence the combined effect is practically zero. At the surface, however, all the molecules are below it, and there are none above to neutralise the force which they exert. There is thus a strong downward force tending to drag the molecule into the surface. This force makes itself manifest in the phenomenon known as surface tension or capillarity.

A molecule will thus require a definite and in most cases, a considerable velocity to succeed in escaping from the surface of the liquid. Now the mean velocity of the molecules depends only on the temperature, and is fixed for any given substance when the temperature is known. The actual velocity of any given molecule at a given instant may, however, differ very considerably from this average speed. The molecules in the liquid are constantly bumping against each other, and at every collision one of the molecules will probably gain speed, and the other lose it, and a difference in velocity is thus set up. The law of the distribution of speeds among an infinite number of colliding molecules has been worked out by Maxwell, for the simpler case of a gas, and is known as Maxwell's law. He showed that at any given instant the great majority of the molecules would have velocities not widely removed from the average, but that there would always be a few which, owing to special circumstances, would have velocities very much greater, and others with velocities very much less, than the normal value.

Thus for oxygen at 0° C., in which the mean velocity of the molecules is 425 metres per second, out of every 1,000 molecules of the gas, about half have velocities between 300 and 500 metres per second, but there are 77 with a velocity greater than 700 metres per second, and 30 with a velocity greater than 900 metres per second.

The case of a liquid is much more complicated than that of a gas, but there is no doubt that a similar state of affairs prevails in the liquid state also. Hence, even at temperatures far below the boiling point of the liquid, there will be a few molecules which have temporarily sufficient speed to escape from the attraction of the mass of liquid below them.

Thus even at low temperatures, a certain number of molecules escape every second from the free-surface of the liquid. If the space above the liquid is free there will thus be a constant evaporation going on until all the liquid disappears into vapour.

If, however, the liquid is contained in a closed

vessel, the escaping molecules are unable to get away altogether and after perhaps many reflections at the enclosing walls will sooner or later strike the surface of the liquid and be reabsorbed. The number so returned to the liquid will depend on the number of molecules present in the vapourous state above the liquid. Evaporation will go on until the number of molecules in unit volume of the vapour is so great that the number returned per second is exactly equal to the number which escape per second owing to their high velocity. Thus for every temperature there will be a corresponding vapour pressure which will be constant for that temperature and that pressure. If the temperature is raised, the average speed of the molecules is increased, and more of them will attain a velocity sufficient to enable them to escape from the surface. The vapour pressure thus rises. When the vapour pressure reaches that of the atmosphere the liquid will begin to boil, and if the supply of heat is maintained, will all pass into the state of a gas.

The difference between a liquid and a gas is that in the latter all the molecules are moving with sufficient speed to escape completely from the attraction of their neighbours, and are thus able to avoid those molecular aggregations which take place in the liquid state. On this account the behaviour of a molecule in a gas lends itself much more readily to mathematical analysis than the more complicated phenomena of the liquid and solid states. The kinetic theory of gases has been very completely worked out by Maxwell, Clausius, and their successors. Unfortunately, even with the simplifications thus introduced, the theory needs considerable mathematical ability for its elucidation, and does not lend itself at all readily to non-mathematical treatment.

The pressure which a gas exerts on its boundaries

is made up of the innumerable shocks given to the bounding walls by the many impacts upon them of the molecules of the gas, which are moving as we have seen with considerable speeds. It will obviously depend on the number of molecules present per unit volume, and the speed with which they are moving. The higher the temperature, the faster the molecules move, and the greater, therefore, will be the pressure they exert. It can be shown that this pressure is simply proportional to the temperature, if we begin to measure the latter not from the freezing point of water, but from the absolute zero of temperature, that is, the temperature at which the molecules would all be reduced to rest. This temperature is about 273° below the zero on the Centigrade scale.

Again, if we halve the volume occupied by the gas, we double the number of molecules present in every cubic centimetre, and hence double the pressure which they exert on the enclosing walls. The pressure is thus inversely proportional to the volume occupied by the gas and directly proportional to its absolute temperature; thus, if P is the pressure, V the volume and T the temperature, of a given mass of substance in the gaseous form, $P.V/T$ is a constant. This is the well-known gas equation embodying the laws of Boyle and Charles.

These laws are obeyed approximately by all gases, but perfectly by none. We have assumed in calculating the formula that the molecules exert no attraction upon each other in the gaseous state. This is of course only approximately true. Hence the velocity with which the molecules strike the boundary is always a little smaller than the normal velocity at that temperature, owing to the attraction of the molecules behind it. This effect will obviously be greater if the molecules are closely packed together than if they occupy a large volume. It has been shown that to get the true pressure we must add to the observed pressure an amount equal to a/V^2, where a is constant for a given gas.

Again the actual space in which the molecules are free to move is less than the whole space of the vessel, owing to the fact that the molecules are not mere mathematical points, but have a definite volume of their own. Thus if b is the whole volume occupied by all the molecules in a gas, the actual free space in which they are free to move is not V the volume of the vessel, but $V - b$.

To be accurate, therefore, we should write our gas equation in the form

$$\left(p + \frac{a}{V^2}\right)\left(V - b\right) = R.T,$$

where R is a constant for the gas.

This equation was first given by Van der Waals, and is found to represent the behaviour of real gases with a considerable degree of accuracy. In fact it is even capable of representing with some degree of success the transition state from liquid to gas. It has the further advantage that it offers us a simple method for determining the one molecular dimension with which we are still unacquainted, namely, the radius of the molecule.

By accurate experiments on the behaviour of the gas under varying conditions, we can determine the value of the constant b in any given case. Since one molecule cannot pass between two others if the space between them is less than the diameter of a molecule, the volume b occupied by the molecules in the gas will be somewhat greater than the sum of their individual volumes. If r is the radius of each molecule, and N the total number of molecules present in the gas, a number which we have already determined it can be shown that

$$b = \tfrac{5}{2} \cdot \tfrac{4}{3} \cdot \pi r.^3 N$$

assuming that the molecules are spheres. In compound gases where there are several atoms in the molecule this can hardly be the case, but in the case of monatomic gases where the atom is also the molecule, it seems a probable assumption. Hence knowing N, we can calculate the value of r the radius of the molecule of the gas we are considering. A few of the values so obtained are given in Table IX. It will be seen that they are of the order of 10^{-8} or one hundred millionth of a centimetre. The radius of an electron is 10^{-13} cms., or about one ten thousandth part of this. The size of an electron thus

TABLE IX.

The Radius of the Molecule.

Substance.	From Van der Waals' Equation.	From the Viscosity.	From the Thermal Conductivity.
Hydrogen	$1 \cdot 16 \times 10^{-8}$ cm.	$1 \cdot 23 \times 10^{-8}$ cm.	$1 \cdot 20 \times 10^{-8}$ cm.
Helium .	1·15 ,,	1·09 ,,	—
Nitrogen.	1·76 ,,	1·75 ,,	1·66 ,,
Oxygen .	—	1·70 ,,	1·55 ,,
Argon .	1·43 ,,	1·68 ,,	—
Carbon dioxide .	1·70 ,,	2·09 ,,	2·16 ,,

bears to the size of the molecule in which it moves very much the same relationship that the diameter of our earth bears to the diameter of its orbit round the sun. A molecule is therefore a system very similar in its relative dimensions to the solar system but on an infinitesimal scale.

The kinetic theory affords other methods of estimating the dimensions of a molecule, but their theory is too complicated to be attempted in the space that remains. The results obtained are included for the sake of comparison in Table IX. The second column of figures gives the value of the radius deduced from considerations of the viscosity of the gas; the last column the value deduced from its thermal conductivity. The three values agree among themselves as well as could be expected.

Let us return now for a brief space to our solid again. Solids may be divided into two classes, according as they do or do not conduct electricity. It is also found that conductors of electricity are also good conductors of heat, while electrical insulators are generally very poor conductors of heat. We have by now come to connect electricity with electrons, and hence an electric current is a flow of electrons from a place of high to a place of low potential. We may regard a conductor, then, as a substance containing electrons which are free to move under the action of an electric field, while in non-conductors the electrons are fixed and unable to follow the impulse of the field.

How are these electrons set free? In the first place it may be noticed that the only good conductors of electricity are metallic, that is to say, electro-positive in character, substances which we know from other phenomena readily part with an electron under the slightest provocation. Now in a solid such provocation may well be supplied by the close propinquity of the neighbouring molecules. It is well known that a charged body will attract light uncharged substances. The attraction of a well-rubbed stick of sealing wax for small pieces of paper is generally our first introduction to the science of electricity. The attraction is of course mutual, the force on the charged body being equal to that on the uncharged paper. Hence an electron in one atom is attracted by a neighbouring uncharged atom, and under favourable circumstances, and especially in the case of an atom only too ready to part with its electrons, the attraction may well be sufficient to enable it to make its escape.

That some such assistance is necessary and is so given, is shown by the fact that a metallic vapour is no better a conductor than any other gas at the same temperature. Thus while metallic mercury at its boiling point is an excellent conductor of electricity, the mercury vapour above it conducts little better than the air it displaces, that is practically not at all.

The evolution of electrons by the molecules of a metal is very similar to the evaporation of a liquid into a closed space—a phenomenon we have already described. It will go on until the number of electrons which are attracted by the positively charged residues and so recombine, is equal to the number of electrons which are given off by the neutral molecules in the same time. Thus for a given temperature there will be a definite pressure of electrons per unit volume in the metal, and we may in fact look upon the electrons as a kind of tenuous gas filling up the interspaces between the molecules of the solid. We must regard them as moving about in all directions in the solid with considerable velocities. The principles of thermodynamics would lead us to imagine that when equilibrium is established each of these electrons will have the same average energy as the molecules around it, that is to say, the energy corresponding to the temperature of the body at the time. This is an important point to which we shall return later.

These electrons, if no electric force be acting, will be moving in all directions, so that if we take any cross section of the metal the number of electrons crossing it in one direction will be the same as the number crossing in the opposite direction, and so the total transference of electricity across the section will be zero.

If, however, we apply an electric field to the body there will be a force on each electron urging it in the direction of the field. Thus in addition to the irregular motion due to the heat energy of the body, there will be a steady drift of the electrons as a whole in the direction of the electric force.

Let us consider for the sake of simplicity, a regular conductor such as a uniform wire. Let us apply an electric field X, to the wire along its length. Then the force on every electron is equal to X*e*, where *e* is the charge on an electron, and it can be shown that the average velocity of the electron due to the electric field is K.X*e*, where K is some constant which for the moment we will leave undetermined.

In one second all the electrons which are less than *v* cms. from a given cross section of the conductor will be forced across it under the action of the field. The volume of this space is *v*A if A is the area of cross section, and thus if there are N electrons per unit volume, each carrying a charge *e* the total charge passing through the given cross section will be N*e*.*v*A units per second. But this is by definition the current along the conductor. Hence remembering that the velocity *v* is K.X*e* we have the current *i* is equal to $K.e^2N.A.X$.

If *d* is the length of the wire and E the difference of potential between its ends, the field X in the wire is equal to E/*d*. Hence finally

$$i = (K \, . \, e^2 \, N) \frac{A}{d} \, . \, E.$$

The current through a given conductor is thus directly proportional to the difference of potential between its ends which is the well known law of Ohm. We see that it is also proportional to the cross section of the conductor, and inversely proportional to its length. We may thus write it

$$i = \sigma.(A/d).E,$$

where σ is written for the quantity in the bracket.

This is a well-known expression for the current through a wire. The quantity σ which is a constant for a given substance at a given temperature, is known as its specific electrical conductivity. Comparing the equations we see that the conductivity σ is equal to $K.e^2N$.

Now the constant K can be evaluated on certain probable assumptions and has been shown to be equal to $V\lambda/6RT$, where V is the velocity of the turbulent motion of the electrons, λ is the average distance traversed by the electron between one collision with a molecule and the next, R is the well known constant from the gas equation, and T is the temperature of the conductor on the absolute scale. Hence finally the electrical conductivity of a substance is given by the expression

$$\frac{N\,e^2\,\lambda\,V}{6.R.T.}$$

This relation contains several quantities which are at present unknown to us, and so we cannot put it directly to the test. Let us proceed, however, a little further.

Suppose now that instead of applying an electric field to the metal, we begin to heat one end of it. We have seen that the turbulent velocity of the electrons depends only on the temperature, and hence the electrons in contact with the heated area will begin to move with a greater speed than those in the colder parts of the substance. These electrons are moving in all directions, and so in course of time, some will move away into the colder regions, carrying their increased velocity with them. Their places near the hot end will of course be taken by electrons from the colder parts which in turn will have their motion accelerated by contact with the hot face.

Thus there will be a constant stream of quickly

moving electrons from the hot end into the cold parts of the body, sharing their energy with, and thus increasing the velocity of, the molecules there. But the temperature at any point in the body depends only on the speed with which the particles are moving. Thus to translate the process from molecular to molar physics, heat has been conducted from the hot to the cold parts of the body, and the rate at which this process takes place measures what is called the thermal conductivity of the substance.

Since then the free electrons play the principal *rôle* both in the conduction of electricity, and the conduction of heat we should expect to find some sort of relation between the conductivity of a substance for heat and its conductivity for electricity.

It has long been known that good conductors of the one are also good conductors of the other, and as long ago as the middle of last century, Wiedemann and Franz, by comparing the experimental results for a number of metals, discovered that the ratio of the thermal to the electrical conductivity had the same value for all metallic substances. Somewhat later Lorenz, by making the comparison at different temperatures, found that the magnitude of this ratio was simply proportional to the temperature at which the comparison was made, reckoned from the absolute zero of temperature. These results are well brought out in Table X. The second column gives the ratio of the thermal to the electrical conductivities at 18° C. for the metals given in the first column of the table. The last column gives the observed temperature coefficient of the ratio. If Lorenz' law is true, this coefficient should have the value ·00366.

For many years this empirical law of Wiedemann and Franz remained without explanation. The electron theory of conduction would, however, as we have seen, lead us to expect with confidence some such relationship. We can, however, go further than this. We have already seen how to calculate the electrical conductivity of a metal. Its thermal conductivity can also be calculated if we assume that the heat is all carried by electrons. This assumption cannot be strictly true, because substances which are non-conductors of electricity still conduct heat to a small but finite degree. In this case the transference of heat is carried on by the molecules themselves, the heated molecules transferring their increased velocity to their immediate neighbours by the somewhat tedious process of repeatedly bumping against them. This method is necessarily slow compared to that of the freely moving electrons, and the very small thermal conductivities possessed by electrical insulators leads us confidently to suppose that the amount of heat transmitted in this way in a metallic conductor is very small, compared with that carried by the electrons.

TABLE X.

Metal.	Ratio of Thermal to Electrical Conductivity.	Temperature Coefficient of the Ratio.
Copper	$6{\cdot}71 \times 10^{10}$	·0039
Silver	$6{\cdot}86 \times 10^{10}$	·0037
Gold	$7{\cdot}09 \times 10^{10}$	·0037
Nickel	$6{\cdot}99 \times 10^{10}$	·0039
Zinc	$6{\cdot}72 \times 10^{10}$	·0038
Cadmium . . .	$7{\cdot}06 \times 10^{10}$	·0037
Lead	$7{\cdot}15 \times 10^{10}$	·0040
Tin	$7{\cdot}35 \times 10^{10}$	·0034
Platinum . . .	$7{\cdot}53 \times 10^{10}$	·0046

The free electrons in a metal behave, as we have seen, like a rarefied gas. We can therefore apply the results obtained for gases to them. The thermal conductivity of a gas has been worked out by Maxwell and others. The mathematical analysis of the problem would lead us too far. It is contained in all treatises on the dynamical theory of gases. It is found that the thermal conductivity of a gas is represented by

$$\tfrac{1}{2}\mathrm{N}\,\lambda\,\mathrm{V.R}$$

where N, λ, V, and R have the same meaning as in the equation for electrical conductivity.

Hence comparing the two equations we have—

$$\frac{\text{Thermal conductivity}}{\text{Electrical conductivity}} = \tfrac{1}{2}\,(\mathrm{N}\,\lambda\,\mathrm{V}\,\mathrm{R}) \div \frac{\mathrm{N}e^2\,\lambda\,\mathrm{V}}{6\,\mathrm{R.T}}$$

$$= 3\,\frac{\mathrm{R}^2}{e^2}\,.\,\mathrm{T}.$$

This is a very simple relation indeed. All the quantities with which we are unfamiliar have gone out, leaving only R the constant of the gas equation, *e* the ubiquitous electronic charge, and T the temperature on the absolute scale at which our comparison is made. Let us substitute for these symbols their numerical values, and so bring our theory to a critical numerical test. R in absolute units is equal to $1{\cdot}40 \times 10^{-16}$, while *e* in electro-magnetic units is $1{\cdot}57 \times 10^{-20}$. If the comparison is made at 18° C., as in Table X., T is (18° + 273°) or 291° on the absolute scale.

Hence, the ratio of the two conductivities should be equal to—

$$3 \times \frac{(1{\cdot}40 \times 10^{-16})^2}{(1{\cdot}57 \times 10^{-20})^2} \times 291,$$

or

$$6{\cdot}95 \times 10^{10}.$$

The agreement with the experimental values given in Table X. is satisfactory beyond the bounds of expectation. Since R and *e* are both independent of the temperature it is evident from our equation, that the ratio varies directly as the absolute temperature as was discovered experimentally by Lorenz.

The electron theory of conduction has many other triumphs and at present also not a few unsolved difficulties, which would need a treatise rather than a chapter to expound. Let us, therefore, leave the subject here, and take a final and farewell glance at our atom, this time in the throes of dissolution.

CHAPTER IX.

THE ATOM IN DISSOLUTION.

IF the ideas which have been developed in the preceding chapters have any sort of foundation in fact, the title of the present chapter will hardly come as a shock to any reader. It may seem at first sight much more surprising that a system so complex as the one we have described, and in such constant motion, when subjected to all the shocks mechanical, chemical, and electrical, which an atom must undergo, should continue to exist in such a persistent and unchangeable way as is actually found to be the case.

The cause is probably to be found in the relatively enormous amount of energy associated with an atom, an amount which is quite out of proportion to the energy liberated by chemical reactions. It has been shown that the heat liberated by the transformation of one gram of radium into its final product is about 4×10^9 gram calories. Compared with this enormous quantity the heat liberated in the most energetic of chemical reactions fades into insignificance. The formation of a gram of water for example, is attended by the emission of about 4×10^3 gram calories. And yet this enormous quantity must only be a small fraction of the total energy in the radium atom. Radium, as we shall see, has an atomic weight of 225 ; the final product of its decomposition one of about 206. This atom, whatever it may be (modern speculation points to lead, but the proof is far from complete) is stable and must possess a very large amount of energy of its own. The total energy in an atom must therefore be enormous, and it is for this reason that all efforts to transmute one element into another have failed.

Great as this energy is, however, our theory indicates that there is a steady and continuous drain upon it which sooner or later must bring about instability, and the dissolution of the system. We saw in the last chapter that an electron in motion round a circle is a source of radiation. The rate of radiation depends on several factors. A single electron moving in a circle is a very efficient radiator, and would quickly drain any atom of its enormous supply. It can be shown, however, that if several electrons are moving round the same circle the amount of radiation is very considerably reduced, so that for a continuous ring of electrons the effect would be very small indeed. Small or great, however, it is a constant source of loss which in the course of ages must, in the absence of some counteracting force with which we are at present unacquainted, bring about the downfall of the atom.

It is doubtful how far this deduction and the theory on which it is based would have found any acceptance had it not been for the discovery, almost simultaneously with the rise of the electron theory, of a series of phenomena which found in it their only possible explanation.

The main outlines of the new science of radio-activity are perhaps familiar to all readers. The subject is one of profound interest, but it lies almost wholly outside the scope of such a work as this. We will, however, summarise such of the main facts of that science as bear on our immediate subject, leaving the proof of the various statements and the description of the often very beautiful experiments from which they were deduced to other hands.

The metal uranium and its compounds were found by Becquerel to give off continuously particles capable of affecting a photographic plate at a considerable distance. These are the β-rays which, as was mentioned in an earlier chapter, are identical except in speed with the cathode rays.

If the uranium salt is subjected to certain chemical processes it is possible to separate the uranium compound into two fractions. In one fraction, an almost imperceptible part of the whole, all the power of emitting β-rays is concentrated. The other, containing practically all the original salt, is completely inactive. For example, if we follow Sir William Crookes and shake up crystals of uranium nitrate in ether, the water of crystallization, which settles out beneath the ethereal solution and can be separated from it by a separating funnel, is found to contain by far the greater part of the β-ray activity of the whole, the great mass of the salt now dissolved in the ether being comparatively inactive.

Let us take these two solutions and keep them under observation for some time. It will be noticed that the watery solution is gradually losing its β-ray producing powers, the rate of emission of these rays falling in fact to half its original value in about twenty-five days. At the same time the ethereal solution which we had rendered inactive, gradually begins to radiate again, so that at the end of twenty-five days it has recovered one-half of its original activity. Fig. 29 shows in a graphic way the exact correlation between the loss of radio-activity by the active fraction and the growth of activity in the inactive mass. It will be seen that the total activity is constant throughout the experiment. From these results we may draw the following conclusions.

In the first place, the property of ejecting these electrons with enormous speeds resides not in the uranium itself, but in some substance which is mixed with it but which can be separated from it by suitable chemical processes. Secondly, this substance is not a stable substance but loses its activity

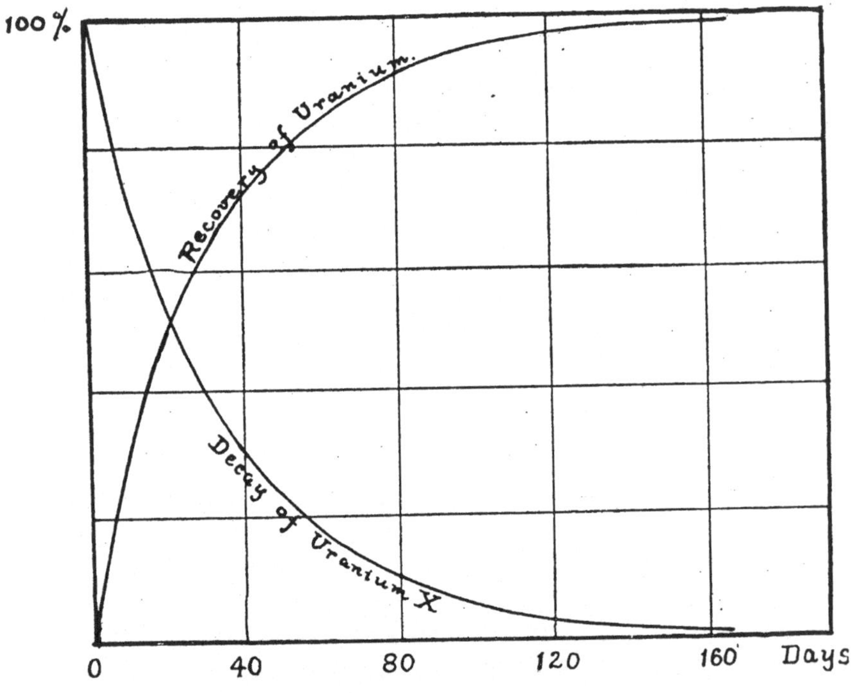

FIG. 29.—Growth and decay of β-ray activity with time for uranium and uranium X.

with time. Lastly, this substance is not of the nature of an impurity in the uranium, but is so vitally bound up with it that when the latter is completely freed from its presence it can reproduce this strange substance again at a steady and perfectly definite rate. The conclusion seems irresistible.

Some new substance is being produced out of uranium itself, and in turn is decomposing with considerable celerity into some other substance which for the moment we are unable to pursue.

I have discussed the case of uranium because that metal is a recognised element belonging to a well-known and respectable family, and might have been expected to behave in a rational and orthodox manner. The case of radium (which is actually a decomposition product of uranium) is still more striking. Rutherford discovered that radium and its salts are constantly giving off a heavy, chemically inert gas known as the "emanation," a non-committal name bestowed upon it before its nature had been determined. This emanation collects in the radium, but can be extracted from it by heating under reduced pressure or better still, by dissolving the salt in water and boiling the solution. In either way the radium can be completely freed from this new gas. On allowing the gas and the radium to stand, however, it is found that fresh emanation makes its appearance in the radium, while the collected emanation gradually disappears, the relation between the growth and decay of the emanation being precisely similar to that of uranium and its β-ray producing product. It is impossible to escape the conclusion that the element radium is evolving out of its own substance the heavy gas known as the emanation.

The properties of both radium and its emanation are now quite well known. Radium is a metal of the calcium group, with an atomic weight of 226·5. It has been prepared in a metallic state, and its properties are found to agree with the place we have assigned to it in the periodic table. The emanation on the other hand is, as we have said, a chemically inert gas resembling the members of the helium group. It can be liquefied at $-150°$ C. Sir William

Ramsay has even succeeded in determining its density, an achievement which as a triumph of skill and refinement of experimental method is beyond all praise. The difficulties of the work will be realised when it is mentioned that the total volume

TABLE XI.

Element.		Average Llfe.	Rays emltted.
Uranium	238·5	6×10^9 years	α, α
Uranium X	(230·5)	24·6 days	β, γ
Ionium	230·5	200,000 years	
Radium	226·5	2,000 years	
Emanation	223 (222·5)	3·85 days	
Radium A	(218·5)	·3 minutes	
Radium B	(214·5)	26·8 minutes	none
Radium C	(214·5)	19·5 minutes	α, β, γ
Radium D	(210·5)	40 years	none
Radium E	(210·5)	4·8 days	β
Polonium	(210·5)	140 days	
?	(206·5)	—	—

of emanation available for the determination was less than one-tenth of a cubic millimetre.

The molecular weight of the emanation deduced from this determination is 223. Assuming the gas to be monatomic like the gases it most resembles, this is also its atomic weight.

What happens to that fragment of the radium

atom which represents the difference between 226·5, the atomic weight of radium, and 223, that of its emanation? The answer is not far to seek. It will be remembered that radium is constantly giving off what are known as α-rays which, as we saw in an earlier chapter, have been proved to be helium atoms with a high velocity and an electric charge. The atomic weight of helium is 4·0. Within the limits of experimental error this is the difference between the weight of an atom of radium and that of the emanation.

The break up of the radium atom therefore takes place as follows. Its energy content becomes too small for its mass and a small fragment of it which we must regard as having been whirling round in the atom with a great velocity, is projected like a stone from a sling, or rather, like one of the fragments of a bursting flywheel. This fragment is always an atom of helium—why, we do not know. The residue of the atom, its atomic weight now reduced by four, forms a new substance, which in this case happens to be gaseous.

The emanation itself is very far from being stable. It is half decomposed in a little less than four days, the new decomposition product being in this case a solid. We cannot follow here the changes undergone by the atom in its attempts to reach a stable state. They are, however, summarised in Table XI., which gives at a glance the whole of the interesting but pitiful story from the break up of the original uranium atom to that of polonium, the last member of the chain to be definitely identified.

The first column gives the name of the element and its probable atomic weight. Those enclosed in brackets have not been actually determined, but are deduced on the assumption that each emission of an α-particle involves a loss of atomic weight of four units. The second column gives the time taken for the element to lose one-half of its mass, or in other words for one-half of its atoms to decompose. It may be called the average life of the atoms of the element. The third column indicates the nature of the rays emitted by the substance on its decomposition.

Polonium itself is radio-active, but its decomposition product has not yet been identified. The worn out atom has probably found peace at last, and hence no longer emits those radiations which in most cases afford our only means of following its changes. It will be seen from the table that five α-particles are given out in the passage from radium to this final product. If each of these is a helium atom the atomic weight of the final element should be $226{\cdot}5 - (5 \times 4)$ or $206{\cdot}5$. This is very approximately the atomic weight of lead, and it has been conjectured with some degree of probability that lead is the final stable product of this great series of changes. At present, however, the evidence can hardly be said to amount to a proof.

Most of these radio-active changes, as they are called, result in the expulsion of an α-particle, and consequent change in atomic weight. Others, however, for example, that of Uranium X, the decomposition product of uranium with which we commenced our survey of the subject, give off only β-particles, and hence differ inappreciably in atomic weight from their immediate successor. The γ-rays which always accompany β-radiation are merely electro-magnetic pulses due to the sudden ejection of the negatively charged particles, and do not concern us here. In some cases, there is apparently no emission of any kind of radiation by the atom, the change in this case consisting of a mere rearrangement of the electrons within the atom itself.

So far we have been treading on safe ground.

There can be no doubt that the scheme we have outlined so briefly above represents the true relation between the various radio-active substances. The disintegration theory which is due to the genius of Professor Rutherford, as much of the work embodied in the table is due to his experimental skill and energy, is the only theory which can in any way explain this novel series of facts. It is now universally accepted. It is when we come to consider the mechanism of the change that difficulties arise.

Of actual knowledge we can hardly be said to have any, but something may be conjectured. We saw that the rings of electrons of which we built up the atom were in many cases only stable if they were revolving round the atom with a velocity exceeding a certain critical value. Owing to radiation from the system the velocity of the electrons must be gradually but constantly decreasing, and a time must come when it will be insufficient to ensure stability.

It seems probable that when this stage is reached one of the electrons leaves the ring and commences an orbit on its own account. The radiation from a single electron, is however, very much greater than from a number of electrons moving in the same orbit, and the loss of energy from the atom is greatly increased. This brings about a state of instability which results in the expulsion of an α-particle or helium atom, though why the expulsion should always take this form is a question we are far from having solved. In some cases the breaking off of this positive portion enables the atomic forces to expel the disturbing β-particle, in which case there is at any rate a temporary lull in the proceedings. In other cases, as in that which follows the break up of the radium atom, four helium atoms are ejected in quick succession, one of the stages lasting only three minutes, until finally radium C expels the disturbing electron, and the succeeding element radium D enjoys an average life of forty years.

Whether this account is correct or no, there is distinct evidence that it is the β-particle which is the disturber of the atomic peace for in every case the expulsion of a β-particle seems to bring about an increase in the stability of the system, the succeeding change having a much longer half period than the changes immediately preceding it.

The subject is one full of difficulties, but where so much has been done we may well expect still more. We are marching with a triumphant army, and are not inclined to set any bounds on the extent of our possible conquests in the future.

In addition to the radium-uranium series, two other elements, thorium and actinium are strongly radio-active, giving rise to series of decomposition products, similar to the chain we have already considered. The question now arises, are all the elements suffering from this slow decay, or are these few heavy elements alone in their spontaneous decomposition ?

Extensive researches have been made to answer this question, but the results are perhaps not quite as definite as might be desired. Potassium has been found to give off appreciable quantities of β-radiation, its activity in this respect being about 1% of that of uranium. Its radio-activity, if such it is, seems to be of a peculiar kind, as in spite of very prolonged and drastic experiments, both by Campbell who discovered the phenomenon, and by others, it has not been found possible to separate from the potassium any substance bearing the same relation to the potassium as Uranium X bears to uranium. Rubidium and possibly cæsium show similar effects.

With all other elements the effect, if it exists, is very small. There is, however, considerable evidence for the radio-activity of all kinds of matter, although it hardly amounts to a formal proof.

If we examine a quantity of air enclosed in a metal vessel we find that, even in the absence of any specific ionizing agent a certain number of ions are always being formed in the air, which enable us to send a small but measurable current through the air. It has been shown that a great part of this ionization is due to radium which disseminated through the atmosphere as emanation, or present in solid form in the soil, seems to be quite ubiquitous. It is found, however, that the amount of this ionization depends on the substance of which the walls of the vessel are made, it is for example, greater in a lead than in a copper vessel. This would certainly seem to point to the emission from the walls themselves of an ionising radiation of some sort, which is greater in amount for lead than for copper.

The matter is complicated by the fact that a certain amount of secondary radiation is set up when radium rays impinge on various metals, and these might very well depend for their intensity on the nature of the metal used. Dr. N. R. Campbell who has made this subject his own, decides against this explanation. I cannot do better than quote his deliberate conclusions.

"It has been proved beyond doubt that the emission of ionizing radiation is an inherent property of all the metals investigated. It is not of course necessary that this ray emission should be identified at once with radio-activity—if that word is taken to mean a process of ray emission accompanied by atomic decomposition. But the constant intensity of the rays, and the probability that the larger portion of them are α-rays, afford considerable support for that hypothesis; while I know of no other process which affords any analogy."

The process is not proceeding at such a rate as to give cause for legitimate alarm. As we have seen, the average life of a uranium atom is about six thousand million years; from a comparison of the relative intensities of the radiation from uranium and from ordinary materials we may conclude that the average existence of the latter cannot be less than ten thousand times longer still. If, however, we accept Dr. Campbell's conclusion that all substances are emitting α-particles then, whether quickly or slowly, all matter is in the process of gradual dissolution into helium, and perhaps hydrogen.

It is curious that in this case the conclusions to be drawn from a sister science, that of astronomical physics, diametrically contradict those which we have arrived at from a consideration of radioactivity. It is a well-known fact that the very hottest and therefore presumably the newest stars, when examined by the spectroscope are seen to consist of very little else than hydrogen, helium, and two other elements, the spectra of which have not yet been identified on the earth. Coming to some of the slightly cooler stars fresh elements all of low atomic weight gradually make their appearance, the list increasing, and elements of higher and higher atomic weight coming into existence as cooler and cooler stars are examined. As we have no reason whatever to suppose that the universe was anything save homogenous in the beginning, we are led to the conclusion that all the elements as we have them on this earth are in reality aggregates of hydrogen, helium, and these two unknown elements.

In these stars progress is obviously from the simple to the complex. Little by little new elements are created by the gradual aggregation of the lighter atoms. On earth, so far as we can follow it the change is in the opposite direction, the heavier atoms such as uranium and thorium spontaneously breaking down into simpler and simpler systems. It must be remembered that so far as we know the decomposition of the radio-active elements is quite uninfluenced by any change of conditions which can be applied to them. The substance can be heated up in the electric furnace or cooled to the temperature of liquid hydrogen without in any way altering its slow but steady rate of decomposition. It is true that the highest temperatures which we can attain are small compared with those of the hotter stars, but it is difficult to conceive that a rise in temperature should assist aggregation, its tendency being always in the direction of disintegration.

The question suggests itself, are the elements as we know them merely a part of a great cycle of growth and decay ? Is the atom born, to grow old, decay, and die ? Are new atoms being formed in the secret places of the universe to take the place of those that have passed away ? The facts have been briefly stated above, but of explanation there appears at present to be none. We are brought not for the first time to one of those mysteries which science has so far failed to illuminate.

We have weighed and measured the atom ; we have analysed it and learned something of the stuff of which it is made and a little of its inner structure, and the way in which it behaves. We have now watched its dissolution and attended its obsequies. Whether it will ever be granted to us to sit as spectators at a new birth time alone can show.

APPENDICES

APPENDIX A.

(*See page* 48.)

DEFLECTION OF THE POSITIVE RAYS BY ELECTROSTATIC AND MAGNETIC FIELDS.

(1) *Electrostatic Deflection.*

Let d be the length of the path of the particles over which the field is applied, and D the distance of the photographic plate Z (Fig. 9) from the centre of the field MN. In the apparatus used d is small compared with D.

Since the field is uniform the particle is acted upon by a force X.E, and acquires an acceleration XE/m in a direction perpendicular to its initial velocity. This acts for the time t taken by the particle to describe the path d, that is for a time d/v. Hence velocity v' acquired under the action of the field

$$=\frac{XE}{m}\,.\,t=\frac{XE}{m}\,.\,\frac{d}{v}.$$

The particle describes the remainder of its path from the field to the plate under these two velocities. Hence the deflection x is to the horizontal distance travelled D as the acquired velocity v' is to the initial velocity v of the particle. Thus

$$\frac{x}{D}=\frac{v'}{v}=\frac{XE}{m}\frac{d}{v^2}$$

$$x = \mathrm{D} \, . \, d \, . \, \frac{\mathrm{XE}}{mv^2}.$$

(2) *Magnetic Deflection.*

The path of the rays in the field is bent into the arc of a circle of radius r where $r = \frac{mv}{\mathrm{H.E}}$;

On leaving the field the particles continue to move along the tangent to this circle at the point where the field ends. Hence the angular deflection θ of the beam is the angle between the tangents to this circle where the path enters and leaves the field respectively. This is obviously the same as the angle between the normals to the circle at those points.

Hence, if d is small compared with r, as is the case in practice, we have

$$\tan\theta = \frac{d}{r} = d \, . \, \frac{\mathrm{HE}}{mv}.$$

But, if y is the deflection as measured on the photographic plate,

$$\frac{y}{\mathrm{D}} = \tan\theta = d \, . \, \frac{\mathrm{HE}}{mv}.$$

Hence

$$y = \mathrm{D} \, . \, d \, . \, \frac{\mathrm{H.E}}{mv}.$$

APPENDIX B.

(*See page* 69.)

Electro-magnetic Mass of an Electron.

With the electron as centre describe two spheres of radius, r and $r + dr$, and draw two radii making angles θ and $\theta + d\theta$ with the direction of motion. If these radii are supposed rotated about the direction of motion of the electron, they will sweep out a hollow cone, cutting from the sphere of radius r a circular zone, the area of which is

$$2\pi r \sin\theta \,.\, r d\theta.$$

The volume of the space included between the two spheres and the hollow cone is thus

$$2\pi r^2 \sin\theta\, d\theta \,.\, dr.$$

The magnetic field due to the moving electron is obviously constant throughout the annular space so defined and equal to

$$\frac{ev\sin\theta}{r^2}.$$

The energy of the magnetic field in the space is thus

$$\frac{e^2v^2\sin^2\theta}{8\pi r^4} \,.\, 2\pi r^2 \sin\theta\, d\theta \,.\, dr, \text{ or } \frac{e^2v^2\sin^3\theta\, d\theta \,.\, dr}{4r^2}.$$

Hence the total magnetic energy in the space between the two spheres is

$$\int_0^\pi \frac{e^2v^2\sin^3\theta}{4r^2} dr \,.\, d\theta = \frac{e^2v^2 dr}{4r^2} \times 2\int_0^{\frac{\pi}{2}} \sin^3\theta\, d\theta.$$

$$= \frac{1}{3}\frac{e^2v^2}{r^2} dr \,.\, \text{since} \int_0^{\frac{\pi}{2}} \sin^3\theta\, d\theta = \frac{2}{3}.$$

The total energy in the field is therefore the integral of this quantity from the surface of the electron of radius a to infinity.

I.e., total magnetic energy associated with the moving electron

$$= \frac{1}{3} e^2 v^2 \int_a^\infty \frac{dr}{r^2}$$

$$= \frac{e^2 v^2}{3a}.$$

APPENDIX C.

The Zeeman Effect.

The mechanical force on the rotating electron P (Fig. 25, p. 115) is H.*e*.*v*, where H is the magnetic field supposed directed down through the paper, *e* the charge, and *v* the velocity of the particle. If P is positively charged, it will act towards the centre O, thus increasing the force retaining the particle; if negative, it will oppose it. The mechanical force on Q will obviously be in the opposite direction to that on P.

Let f be the retaining force in the absence of any magnetic field. Then, since the motion is simple harmonic, we may write

$$f = k.r$$

where r is the radius of the circle described by the particle. Also, by the laws of mechanics,

$$f = \frac{mv^2}{r} \quad . \quad . \quad . \quad (1)$$

If T is the time taken to complete the circle once

$$v = \frac{2\pi r}{T} \quad . \quad . \quad . \quad . \quad (2)$$

$$\therefore \; k.r = \frac{m}{r} . \left(\frac{4\pi^2 r^2}{T^2}\right) \qquad k = \frac{4\pi^2 m}{T^2} \qquad (3)$$

Now, if T_1, r_1, v_1 are the values of T, r, and v for P when the field is applied, and T_2, r_2, v_2 the corresponding values for Q, we have

$$\frac{mv_1^2}{r_1} = kr_1 + \mathrm{H}ev_1 \quad \text{for P}$$

$$\frac{mv_2^2}{r_2} = kr_2 - \mathrm{H}ev_2 \quad \text{for Q}$$

where the charge e is supposed to carry its own sign.

Substituting in these equations for the velocities v_1 and v_2 from (2), and for k from (3), and dividing the equations by r_1 and r_2 respectively, we have

$$\frac{4\pi^2 m}{T_1^2} = \frac{4\pi^2 m}{T^2} + \mathrm{H}e \cdot \frac{2\pi}{T_1} \quad . \quad . \quad (4)$$

$$\frac{4\pi^2 m}{T_2^2} = \frac{4\pi^2 m}{T^2} - \mathrm{H}e \cdot \frac{2\pi}{T_2} \quad . \quad . \quad (5)$$

Subtracting (5) from (4)

$$4\pi^2 m\left\{\frac{1}{T_1^2} - \frac{1}{T_2^2}\right\} = 2\pi \cdot \mathrm{H}e\left\{\frac{1}{T_1} + \frac{1}{T_2}\right\}$$

$$2\pi m\left\{\frac{1}{T_1} - \frac{1}{T_2}\right\} = \mathrm{H}.e$$

$$\frac{e}{m} = \frac{2\pi}{\mathrm{H}} \cdot \frac{T_2 - T_1}{T_1 T_2}$$

or writing T_0^2 for $T_1.T_2$, which is permissible since the change in the periods produced by the magnetic field is very small and T_0 lies between T_1 and T_2,

$$\frac{e}{m} = \frac{2\pi}{\mathrm{H}} \frac{T_2 - T_1}{T_0^2}$$

But if λ is the wave length of the light emitted, and V is the velocity of light,

$$\lambda = \mathrm{VT}.$$

Thus, if λ_0 is the original unaltered wave length,

λ_c that of the clockwise rotation, λ_a that of the counter-clockwise rotation,

$$\frac{e}{m} = \frac{2\pi}{\mathrm{H}} \cdot \frac{\frac{\lambda_c}{\mathrm{V}} - \frac{\lambda_a}{\mathrm{V}}}{\frac{\lambda_0^2}{\mathrm{V}^2}}$$

$$= \frac{2\pi}{\mathrm{H}} \cdot \frac{\lambda_c - \lambda_a}{\lambda_0^2} \cdot \mathrm{V}.$$

Thus, if the particles are positively charged, the clockwise rotation will have the greater wave length; if negative, the counter-clockwise, as viewed along the lines of magnetic force. The latter is found experimentally to be the case.

TABLE OF ATOMIC DATA.

(Calculated from the most recent observations.)

Mass of an electron m	$8\ 99 \times 10^{-28}$ gm.
Charge on an electron e or monovalent ion	$4{\cdot}77 \times 10^{-10}$ e s. units $1{\cdot}59 \times 10^{-20}$ e m. units
Ratio of charge to mass e/m for an electron	$1{\cdot}774 \times 10^{-7}$ e.m. units per gm.
Radius of an electron	$1{\cdot}87 \times 10^{-13}$ cm.
Electrochemical equivalent of hydrogen (Mass/charge)	$1{\cdot}04 \times 10^{-4}$ e.m. units per gm.
Mass of a hydrogen atom	$1{\cdot}66 \times 10^{-24}$ gm.
Radius of hydrogen atom	$1{\cdot}21 \times 10^{-8}$ cm.
Gaseous molecules in 1 c.c. at 0° C. and 760 mm. pressure	$2{\cdot}705 \times 10^{+19}$
"Gas Constant," R	$1{\cdot}40 \times 10^{-16}$ ergs. per degree
Velocity of light, V	$2{\cdot}998 \times 10^{+10}$ cm. per sec.

BIBLIOGRAPHY

THE following brief list of books and scientific papers may be found of use by readers wishing to pursue further the subjects treated of in the foregoing pages. It lays no claim to completeness.

BOOKS.

Professor Sir J. J. Thomson's *Conduction of Electricity through Gases*, and Professor Sir E. Rutherford's *Radioactivity* are the standard works on the physics of the electron and on radioactivity disintegration respectively, while Dr. N. Campbell's *Modern Electrical Theory* is a fascinating and original treatise on the electron theory as a whole. For the older kinetic theory of matter, students cannot do better than consult Professor O. E. Meyer's treatise.

The above books are of an advanced character and postulate some considerable mathematical and physical knowledge on the part of their readers. Of books of a more popular type Professor Sir J. J. Thomson's *Electricity and Matter*, or his more recent *Corpuscular Theory of Matter* may be consulted, while for a detailed account of the experiments described in Chapter IV. the same author's monograph on *Positive Rays* should be referred to.

SCIENTIFIC MEMOIRS.

For those who may wish to consult the original authorities a list of a few of the most important papers is appended.

Sir J. J. THOMSON. "Kathode Rays." *Phil. Mag.*, October, 1897. "The Charge of Electricity carried by the Ions produced by Röntgen Rays." *Phil. Mag.*, Vol. xlvi., 1898.

R. A. MILLIKAN. "Charge on an Electron." *Physical Review*, 1911. "Elementary Electric Charge." *Physical Review*, p. 109, August, 1913.

C. T. R. WILSON. "The Photography of Particles ejected from Atoms." Proceedings of Royal Society (*Proc. Roy. Soc.*), A. 1913.

E. RUTHERFORD and H. GEIGER. "The Charge and Nature of the α-particle." *Proc. Roy. Soc.*, A. 1908.

Sir J. J. THOMSON. "Rays of Positive Electricity." *Phil. Mag.*, February, 1911. Further experiments, *Phil. Mag.*, August, 1912.

KAUFMANN. "Ratio of Charge to Mass for High Speed Electrons." *Göttingen Nach.*, November, 1901.

Sir J. J. THOMSON. "The Structure of the Atom." *Phil. Mag.*, March, 1904. *Phil. Mag.*, October, 1913. "The Zeeman Effect." *Proc. Camb. Phil. Soc.*, Vol. xxxix.

E. RUTHERFORD and F. SODDY. "Radioactive Change." *Phil. Mag.*, 1903.

Dr. R. W. GRAY and Sir W. RAMSAY. "The Density of Radium Emanation." *Chemical News*, 1911.